Springer Series in Surface Sciences

Volume 69

This series covers the whole spectrum of surface sciences, including structure and dynamics of clean and adsorbate-covered surfaces, thin films, basic surface effects, analytical methods and also the physics and chemistry of interfaces. Written by leading researchers in the field, the books are intended primarily for researchers in academia and industry and for graduate students.

Wiesław Jakubik

Application of Surface Acoustic Waves in Gas Sensors

Theoretical Models and Advanced Sensor Structures

Springer

Wiesław Jakubik
Institute of Physics - Centre for Science
and Education
Silesian University of Technology
Gliwice, Poland

ISSN 0931-5195 ISSN 2198-4743 (electronic)
Springer Series in Surface Sciences
ISBN 978-3-031-89973-7 ISBN 978-3-031-89971-3 (eBook)
https://doi.org/10.1007/978-3-031-89971-3

This Springer imprint is published by the registered company Springer Nature Switzerland AG
The registered company address is: Gewerbestrasse 11, 6330 Cham, Switzerland

The author dedicates this book to his entire family,
and in particular, dear wife Anna,
son Andrew with wife Monica and Asia,
and daughter Eve with husband Paul

Autor dedykuje niniejszą książkę całej swojej rodzinie,
a w szczególności, kochanej żonie Annie,
synowi Andrzejowi z żoną Moniką i Asią,
oraz córce Ewie z mężem Pawłem

Preface

Science is only possible because, you can say something, without knowing everything

—Hermann Bondi

The monograph is the author's presentation of the most important issues related to the application of surface acoustic waves (SAW) in gas sensors and is a revised and expanded version compared to the Polish edition in 2012. More than 30 years' experience in research work with such sensors obliges the author to take up this topic. On the other hand, the conduct of more than tens of thousands of experimental studies, with different types of sensor structures, has allowed to draw concrete conclusions regarding their sensor properties and the possibilities of practical applications.

Some of the research results have been published in several international papers and presented at conferences, including IEEE Sensors (Vienna 2004), ECS-Electro Chemical Society Meeting (Los Angeles 2005), Forum Acusticum (Budapest 2005), Eurosensors (Gothenburg 2006, Lausanne 2009, Athens 2011, Freiburg 2015), Semiconductor Gas Sensors (SGS) (Zakopane 2008, Kraków 2010), SAW Symposium (Vienna 2014, Dresden 2018), and COE (Wrocław 2024). The book contains the most important results, supplementing their interpretation with relevant mathematical models, including the model of acoustoelectric interactions for bilayer sensor structures in the SAW system.

An important novelty in the monograph is the presentation of the possibility of using not only single sensor layers, but above all layered structures in SAW sensors. Sensor structures of this type, consisting of two or more thin layers, make it possible to obtain, in some cases, much higher sensitivity than single layers. A significant increase in sensitivity can also be achieved for single photoconductive polymer structures with additional optical activation, which is the latest research trend.

Sharing the thought of Hermann Bondi, the author expresses the hope that the issues contained in the monograph have been presented in a comprehensible way. In addition, the author thanks all those people from the Institute of Physics and the current Department of Optoelectronics of the Silesian University of Technology, Gliwice, Poland with whom he had the honor to collaborate during many years of painstaking experimental work, while being aware of his own contribution to the development of gas sensors with SAW on a global scale.

Gliwice, Poland
December 2024

Wiesław Jakubik

Acknowledgments The author expresses his gratitude for the opportunity to conduct two research stays abroad in November 2023 (Institute for Photonics and Nanotechnology IFN-CNR and Institute of Microelectronics and Microsystems CNR, Rome, Italy) and in June 2024 (Leibniz Institute for Solid State and Materials Research IFW Dresden, Germany), during which he was inspired, thanks to the creative scientific atmosphere in the field of SAW applications, to write an English and supplemented version of this book.

Both visits were supported by the Rector's professorial grant implemented as part of the Excellence Initiative—Research University Programme—Silesian University of Technology, Gliwice, Poland, grant number 14/030/SDU/10-07-01.

The author also expresses a big thanks to all the people of Springer Nature, who, thanks to their professional support, despite many difficulties on the part of the author for using outdated software, made it possible to create this book.

Competing Interests The author has no competing interests to declare that are relevant to the content of this manuscript.

Contents

1 Introduction ... 1
References ... 4

2 Gas Sensors with SAW ... 5
2.1 The Essence and Features of SAW Sensors ... 6
2.1.1 The Essence of the Operation of SAW Gas Sensors ... 7
2.1.2 Physicochemical Phenomena in SAW Sensors ... 10
2.1.3 Features and Importance of SAW Gas Sensors ... 11
2.2 Generation and Detection of SAW ... 13
2.2.1 Interdigital Transducer ... 13
2.2.2 Delay Line and Resonator ... 15
2.2.3 Oscillator System ... 15
2.3 Types of Sensors with SAW ... 17
2.3.1 Rayleigh-Type SAW Sensors ... 18
2.3.2 Other Types of Acoustic Sensors ... 21
2.4 Summary and Comparison of Acoustic Sensors ... 23
References ... 29

3 Theoretical Models of SAW Sensor's Operation ... 33
3.1 General Model ... 34
3.2 Basic Couplings in Gas Sensors ... 35
3.3 Kinetic Phenomena for the Interaction Gas-Sensing Element ... 37
3.4 Models of Coupling Gas-Sensor Element ... 40
3.4.1 Metal Oxides ... 40
3.4.2 Phthalocyanines ... 42
3.4.3 Thin Palladium Layers ... 44
3.4.4 Layered Structures ... 46
3.4.5 The Change of Sensing Element Parameters ... 47
3.4.6 Response Profiles of SAW Gas Sensors ... 50
3.5 Theoretical Models of Coupling: Sensor Element-SAW ... 52
3.5.1 Types of Couplings ... 53
3.5.2 Perturbation of Complex Propagation Constant ... 55

3.5.3 Perturbations for Mechanical Effects 57
3.5.4 Perturbations for Acoustoelectric Effect 64
3.6 Summing-Up 86
References 89

4 Methodology of SAW Gas Sensor Testing 93
4.1 Preparation of SAW Sensor Module 94
4.2 Measuring Method with Single-Delay Line 97
4.3 Measuring Method with Dual-Delay Line 97
4.4 Method with Frequency Mode Identification 101
4.5 Method with Multi-channel Sensor Modules 103
4.6 Acoustoelectrical Method 105
4.7 Measuring Station 106
4.7.1 Chambers and Measuring Instruments 109
4.7.2 Gas Metering System 110
4.8 Phenomena Related to the Methodology of Testing Sensors with SAW 111
4.8.1 Thermal Drift 111
4.8.2 Frequency Changes 113
4.9 Summary 114
References 114

5 Study of Selected Sensor Structures with SAW 117
5.1 Single Sensor Structures 118
5.1.1 Palladium Layers 118
5.1.2 Analysis of Mass-Elastic Interactions for Pd Film 119
5.1.3 Copper Phthalocyanine Layers (CuPc) 121
5.1.4 Tungsten Oxide (WO_3) Layers 122
5.1.5 Single Layers in NO_2 Detection 123
5.2 Bilayer Structures 124
5.2.1 Phthalocyanine-Palladium Bilayer Structures 124
5.2.2 Bilayer Structures Metal Oxide-Palladium 135
5.2.3 The Use of Bilayer Structures in Hydrogen Detection 137
5.2.4 Bilayer Structure Nafion®-Polyaniline 138
5.3 Multilayered Structures 141
5.3.1 Additional Polyethylene Layers 141
5.3.2 Additional TPX Layer 142
5.4 The Results of the Tests Using the Speed of Interaction 147
5.5 Research Results in Multi-channel System 149
5.6 Acoustoelectrical Studies of Bilayer Structures 152
5.6.1 Summary of Experimental Research 154
References 155

6 SAW Gas Sensors with Light-Activated Sensor Structures 157
6.1 Genesis and Essence of Light-Activated Sensor Structures (LASS) 157
6.2 Model of Acoustoelectric Interactions in LASS 159
6.3 Methodology of LASS Testing 161
6.3.1 Switchable Oscillator System 161
6.3.2 Experimental and Gas Dosing System 161
6.3.3 Light Activation Details 164
6.4 Studies of LASS for DMMP Detection 164
6.4.1 SAW Module with Photoconductive rrP3HT Films 164
6.4.2 Activation by Laser LEDs and LEDs 165
6.4.3 Sensitivity on the Number of Photons 169
6.4.4 Selectivity, LOD, and Response Time 170
6.4.5 Repeatability and Stability of rrP3HT Films 170
6.5 Acoustoelectrical Method in LASS 172
6.5.1 Electrical Planar Transducer 172
6.5.2 Explanation of Sensitivity Increase 174
6.6 Summary for LASS 177
References 178

7 Summing-Up 181
References 184

Index 187

Abbreviations and Symbols

Abbreviations

AE	Acoustoelectric
BAW	Bulk acoustic wave
CCD	Charge coupled devices
CMOS	Complementary metal-oxide-semiconductor
CNT	Carbon nanotubes
CVD	Chemical vapor deposition
DMMP	Dimethylomethylophosphor
FPW	Flexural plate wave
HF	High frequency
HIgG	Human immunoglobulin type G
IDT	Interdigital transducer
IMC	Inverse mode configuration
LA	Light activation
LASS	Light-activated sensor structures
LED	Light emitting diode
$LiNbO_3$	Lithium niobate
LOD	Limit of detection
MEMS	Micro-electromechanical systems
MIC	Microwave integrated circuits
MOS	Metal-oxide-semiconductor
MPc	Metalophthalocyanines
MWCNT	Multi-walled carbon nanotubes
NMC	Normal mode configuration
OVG	Owlstone vapor generator
PC	Photoconductive properties
Pd	Palladium
p.s.e	Parameters of the sensor element
PPAa	Plasma polymerized allylamine
PVD	Physical vapor deposition

QCM	Quartz crystal microbalance
RH, rh	Relative humidity
rrP3HT	Regioregular poly-3-hexylothiophene
RSI	Relative sensitivity increase parameter
SAW	Surface acoustic waves
SH-APM	Shear-horizontal acoustic plate mode
SilPEG1.4	Kind of photoconductive polymer
SPUDT	Single-phase unidirectional transducer
SSBW	Surface skimming bulk wave
STW	Surface transverse wave
SWCNT	Single-walled carbon nanotube
swp	Structure work point
UV	Ultraviolet radiation
WO_3	Tungsten oxide

Symbols

A	Amplitude of oscillator signal, wave amplitude, amplifier symbol
A_p	Constant depending on the parameters of the piezoelectric substrate material
A	Aperture of interdigital transducer, width of the acoustic beam
a_z	Additional external acceleration
α	SAW damping on free substrate or angle of light illumination
α'	SAW damping on substrate with sensor structure
$\Delta\alpha$	SAW damping change
B	Coefficient related to the relative rate of adsorption and desorption
β_j	Constant propagation of partial waves generated to the sensor structure
c	Concentration of vapor or gas in the surrounding atmosphere
c_0	Gas concentration in the measuring chamber at steady state
c_{se}	Concentration of substance in the sensor element
$\mathrm{c_j}$	Coupling parameters between SAW and piezosubstrate
C_s	Capacity per unit length of pair of interdigital transducers
C_{11}, C_{44}	Elastic modulus of the sensor element
d	Substrate thickness
D_0	Hydrogen diffusion ratio to palladium volume
D_j	Component vector of electric induction
D_e	Constant diffusion of electric charges
E	Electric field strength
E_z	Strength of the external electric field
$E^{(j)}$	Strain modulus taking into account the relative stress-to-strain ratio
ε	Semiconductor electrical permeability
ε_0	Vacuum electric permeability

ε_p	Electrical permeability of the substrate
ε_p^T	Effective electrical permeability of the piezoelectric substrate at a constant stress field T
f	Wave frequency, line frequency with sensor structure
f_0	Frequency of oscillation on the free surface of the substrate (without cover)
Δf	Differential frequency
η	Dynamic viscosity of the viscoelastic structure
ϕ_A	Amplifier phase shift
ϕ_{tr}	Single transducer phase shift
φ	Electric potential
φ_c	Total electric potential
$\Delta\varphi$	Difference in electric potentials as a result of dipole layer production
G'	Real part of sensor element material shearing module
G''	Imaginary part of sensor element material shearing module
h	Thickness of the sensor layer
h_1	Thickness of the first film in bilayer structure
h'	Planck constant
ΔH_g	Change in the entropy as a result of gas sorption
I	Intensity of light illumination
J	Current density in the semiconductor layer
k_0	SAW wave number
k	Combined SAW propagation constant without sensor structure
k'	Combined constant of SAW propagation after the sensor structure is made
Δk	Change of the combined SAW propagation constant
k_B	Boltzmann constant
K	Partition coefficient
K'	Real part of the sensor layer volumetric module
K''	Imaginary part of the volumetric module of the sensor layer
K^2	Electromechanical coupling coefficient
κ	Delay line coverage factor
l	Coverage width of the electrode system in the planar electrical method
L	Length of the delay line
L_s	Length of sensor structure
λ	Wavelength of the SAW or light illumination
λ'	Lame constant of the isotropic sensor layer
λ_D	Debye shielding length
m	Mass of a gas molecule
Δm	Change in the surface density of the sensor element
M_j	Elastic modules for acoustically thick structures
μ'	Lame constant or isotropic sensor layer
μ_H	Dipole moment of adsorbed hydrogen atom
μ	Mobility of electric charge carriers
n	Number of SAW wavelengths located along the length of the structure

n_0	Number of SAW wavelengths between transceiver and receiver
N	Number of electrode pairs of interdigital transducer
N_s	Density of surface active centers (absorption states)
p	Width of the electrode and gap of the interdigital transducer
p'	Particle pressure of gas particles in the surrounding atmosphere
p_z	External pressure
P	Average flux of wave power per unit of sound beam width
Q	Gas flow through the measuring chamber
Q_z	External flow
R	Gas constant
R^*	Parameter defining the boundaries between acoustically thin and thick structures
R_{sam}	Sample electrical resistance determined from measurements
R_{s0}	Sample resistance after deposition
$R_s(light)$	Sample electrical resistance under light activation
ρ_0	Initial surface mass density of the sensor structure
$\Delta\rho$	Change in the surface mass density of the sensor element
ρ_e	Spatial density of electric charges
ρ_p	Density of substrate
ρ_{srf}	Average surface mass density of the structure
ρ_s	Surface density of electric charges
S_{AE}	Acoustoelectric sensitivity of the single structure/or under the light activation
$S_{\Phi N}$	Sensitivity on the number of photons
S_m	Mass sensitivity
S_m^f	Mass sensitivity in the oscillator system
ΔS_g	Change in entropy due to gas sorption
S_{jz}	Components of deformation tensor
σ	Electrical conductivity of the sensor layer
σ_{rs}	Electrical conductivity of sensor structures obtained from planar measurements
σ_s	Electrical surface conductivity of the sensor layer
σ_{s0}	Initial surface electrical conductivity of the structure
$\sigma_{s(light)}$	Electrical surface conductivity of the polymer film calculated from the sample resistance $R_{s(light)}$ after the light activation
$\Delta\sigma_s$	Change in surface electrical conductivity
t	Time
τ	Time delay of acoustic line
τ_0	Signal delay time on free surface of substrate
τ_K	Measuring chamber filling time constant
τ_r	Residence time (gas molecules remaining on the surface of the structure)
θ	Degree of coverage of sensor element
T_{mz}	External mechanical stress
T_{ij}	Components of the mechanical stress tensor

T_z	External temperature
T	Temperature of action of the sensor structure
u_C	Total wave energy density per unit area
$\boldsymbol{u}$	Field of displacement of substrate particles around equilibrium positions
ω	Wave circular frequency
ω_0	Circular frequency (pulsation) of undisturbed wave
ξ	Acoustoelectric parameter of single structure
ξ_1	Standardized electrical conductivity of the first layer in bilayer structure
ξ_2	Normalized electrical conductivity of the second layer
$\xi_{light}\xi(\lambda)$	Acoustoelectric parameter of the sensor structure under the light activation/under the light with specific wavelength
v	Phase speed of SAW propagation
v_0	SAW speed on the free surface of the piezoelectric crystal
v'_0	SAW speed after the sensor element manufacturing
Δv	SAW speed change
v_j	Velocity components of points on the surface of the substrate
x	Design parameter determining the electrical conductivity ratio of the bilayers
y	Distance from the surface of the SAW propagation
z	Direction of the SAW propagation
$z_E{}'$	Normalized electrical surface impedance of the layer
Z_{mj}	Surface mechanical impedance of the sensor structure
Z_0	Electrical impedance

Chapter 1
Introduction

Abstract This chapter provides an introduction to the application of gas sensors in particular surface acoustic wave sensors with emphasis on their unique characteristics related to high mass sensitivity and the possibilities of acoustoelectric interactions. In addition, the contents of the individual chapters are presented.

Continuous industrial and technological development and the increasing needs for safety and environmental protection make it necessary to precisely monitor the content of various substances in air, water, and soil.

In many cases, performing accurate analyses, often of very small amounts, e.g. toxic, explosive, flammable gases in the air or heavy metals in water, involves the use of expensive stationary laboratory methods such as gas chromatography, mass spectroscopy, or infrared spectrometry. Although monitoring of this type is extremely accurate, new ways are being sought that enable analyses to be performed not only in the laboratory, but also outdoors, at the site of contamination or hazard. One such way is to design all kinds of sensors for many substances and chemical compounds in the gas, liquid, or even solid phase.

Generally, a sensor is a device that is able to process certain non-electrical quantities into a convenient electrical signal or other physical "signal" (e.g. acoustic—change of frequency or phase, gravitational—mass change, optical—refraction index perturbation, etc.) [3]. A chemical sensor is a device which produces a signal (usually electrical) that is a function of the concentration in its environment of a substance or chemical compound. The sensor system can also be used to characterize sensor elements made of different materials. Therefore, chemical sensors can be used both in basic studies of these materials and can also be used for the detection and measurement of many chemicals [4].

The main characteristics of the sensors are their easy portability, speed of operation, and low price. Therefore, the sensors should be lightweight, have small dimensions, and have a simple design and manufacturing technology. With the sensors, it is not necessary to take samples for analysis as in the laboratory methods. As a result, the total measurement time is significantly reduced, which is extremely important in health and life-threatening situations. The threshold sensitivities obtained, at which

W. Jakubik, *Application of Surface Acoustic Waves in Gas Sensors*, Springer Series in Surface Sciences 69, https://doi.org/10.1007/978-3-031-89971-3_1

the sensor is able to produce a measurable signal, and the measurement accuracy are often comparable or only slightly inferior to time-consuming laboratory methods. On the other hand, it is more difficult to achieve high selectivity, i.e. as much sensitivity as possible to one selected compound or substance. In this respect, laboratory methods, based on other mechanisms of action, are generally better, allowing precise determination of the content of even a few substances simultaneously.

Currently the most famous are sensors using acoustic, electrochemical, optical, resistive, and thermal methods. Each method can be further divided into further groups and subgroups. Among acoustic methods, sensors based on volume waves can be distinguished—bulk acoustic wave (BAW) and surface acoustic waves (SAW). SAW sensors can be divided according to the types of excited surface waves, the piezoelectric substrate used, as well as the target environment (gas and/or liquid). The application of a given type of SAW in a gas or liquid environment is determined by the directions and magnitude of the excited mechanical displacements on the propagation surface, which will be discussed in the next chapter of the book (see Sect. 2.3.2). At present, however, the selection of the test object in the form of the SAW Rayleigh-type gas sensors, together with the corresponding sensor coatings, will be brief justified.

The first gas sensor with Rayleigh's SAW was presented in the paper [6]. It used a quartz piezoelectric substrate and various types of polymer coatings to detect pairs of organic compounds in the air. The most important part of the gas sensor with SAW therefore consists of a piezoelectric substrate, which enables easy excitation of surface waves by means of interdigital transducers (IDTs) (see Sect. 2.2), and a corresponding sensor element produced on its surface, in the area of wave propagation.

The possibilities of using Rayleigh's SAW in gas sensors are primarily due to the extremely high sensitivity of this type of wave to changes in the mechanical and electrical boundary conditions of its propagation. This is the main reason for the high sensitivity of this kind of sensor systems. It is noteworthy that the practical limit of mass detection is estimated at ~250 pg/cm^2 for wave with a frequency of ~100 MHz [5]. The ability to detect changes in mass is the main task of various types of sensor elements with very low (e.g. some polymer coatings) or very high (e.g. metallic layers) electrical conductivity in the area of wave propagation. On the other hand, the use of sensor coatings with appropriately matched electrical conductivity, relative to the parameters of the piezoelectric substrate, makes it possible to use the so-called acoustoelectric sensitivity. The sensor is able to detect changes in the electrical conductivity of semiconductor sensor elements. In some cases, especially for bilayer sensor structures such as semiconductor–metal, the acoustoelectric (AE) sensitivity can significantly exceed the mass sensitivity [1]. In modern SAW gas sensors, a significant increase in AE sensitivity can also be achieved by light activation of appropriate polymeric photoconductive thin layers such as regioregular poly-3-hexylothiophene (rrP3HT) for dimethylomethylophosphor (DMMP) detection [2, 7].

The type of material, the manufacturing technology, and the geometric configuration of the sensor element (e.g. single-layer, bilayer, or even multilayer structure) have a significant impact on obtaining appropriate sensor parameters, i.e. high sensitivity, satisfactory selectivity, or short response and regeneration times. The ability to achieve high sensitivity in SAW gas sensors also results from certain specific properties of the surface waves themselves (see Sect. 2.1).

Chapter 2 presents the general characteristics of SAW sensors, taking into account the essence of the possibility of using this type of waves in gas sensors, as well as the basic method of their generation and detection in piezoelectric substrates by means of interdigital transducers (IDT). The most commonly used types of SAW sensors are also presented, as well as widely recognized characteristics of this type of gas sensors and their importance for sensor applications.

Chapter 3 contains the general models of the SAW sensor operation along with their mathematical description. The general model is based on three basic couplings occurring in gas sensors: gas molecules-sensor structure, then sensor structure-SAW propagation and measurement of altered SAW parameters by a suitably sensitive detection system. The mathematical description takes into account the influence of the most important parameters of the layers and sensor structures on the SAW propagation. This description adopts a typical division into the mass effect, resulting from the mechanical parameters of the sensor structure, and the acoustoelectric effect, which is a consequence of its electrical parameters, mainly electrical surface conductivity. For the first time, a detailed theoretical description of the acoustoelectric effect for bilayer sensor structures was presented.

In Chap. 4, the most important features of the dual delay line measuring system often used in practice are described. This system compensates for the influence of some undesirable external factors on the measurement results, e.g. small changes in temperature or pressure. The system enables precise examination (by measuring frequency changes) of different sensor layers and structures made in the SAW system, placed in different gas atmospheres. In order to confirm the acoustoelectric interactions that result from changes in the electrical conductivity of the sensor layer or structure, the so-called planar electrical method was additionally introduced. It makes it possible to measure changes in the electrical resistance of thin layers and sensor structures under the influence of the presence of different gases. For sensor structures produced in the same technological processes, it is possible to correlate acoustic and electrical responses, and thus confirm the occurrence of the aforementioned acoustoelectric interactions.

Chapter 5 describes the most important experimental work carried out by the author using sensor modules with double-delay line, made on the substrate of Y–Z lithium niobate. This substrate is characterized by a good electromechanical coefficient, which provides in many cases decent acoustoelectric responses. The work concerns the sensory properties of selected single, bilayer structures, as well as prototype multilayer structures, in which additional layers perform limiting functions, e.g. the influence of water vapor. However, the most attention has been paid to the semiconductor–metal bilayer structures, e.g. phthalocyanine-palladium and tungsten oxide-palladium, as they are the most promising in terms of practical applications in hydrogen sensors.

Chapter 6 presents the latest possibilities of SAW applications in gas sensors through prototype light-activated sensor structures. The nature and genesis of such structures are included, along with a qualitative model of acoustoelectric interactions, test methodology with a novel switched-channel oscillator system, and test results for thin regioregular poly-3-hexylothiophene (rrP3HT) photoconductive polymer layers in dimethylomethylophosphor (DMMP) detection (safe simulant of warfare chemical agents such as sarin). Light activation methods such as white light, low-power LED lasers, and LEDs are presented, along with ways of modulating the activation parameters by changing the intensity and wavelength of the light.

The summary highlights the author's original contribution to the development of gas sensors with SAW. The prospects of applications and further research taking into account the latest solutions based on light-activated photoconductive polymer structures are also indicated.

References

1. Jakubik WP, Urbańczyk MW, Kochowski S, Bodzenta J (2002) Bilayer structure for hydrogen detection in a surface acoustic wave sensor system. Sens Actuators B 82(2–3):265–271
2. Jakubik W, Wrotniak J, Kaźmierczak-Bałata A, Stolarczyk A, Powroźnik P (2023) Light-activated SAW sensor structures with photoconductive polymer films for DMMP detection. Sens Actuators B 384:133597. https://doi.org/10.1016/j.snb.2023.133597
3. Teterycz H (2005) Grubowarstwowe chemiczne czujniki gazów na bazie dwutlenku cyny (Thick-layered chemical gas sensors on the base of tin oxide). Politechnika Wrocławska, Wrocław
4. Thomson M, Stone DC (1997) Surface-launched acoustic wave sensors. Wiley, New York
5. Vellekoop MJ (1998) Acoustic wave sensors and their technology. Ultrasonics 36:7–14
6. Wohltjen H, Dessy R (1979) Surface acoustic wave probe for chemical analysis. Anal Chem 51:458–1470
7. Wrotniak J (2022) Akustyczne fale powierzchniowe w badaniach własności sensorowych cienkich warstw wybranych związków polimerowych z wykorzystaniem dodatkowych metod aktywacji (Surface Acoustic Waves in studies of sensor properties of selected polymer compounds thin films with the use of additional activation methods). Dissertation, Silesian University of Technology, Gliwice

Chapter 2
Gas Sensors with SAW

Abstract This chapter presents the general characteristics of SAW sensors, taking into account the essence of the possibility of using this type of waves in gas sensors, as well as the basic method of their generation and detection in piezoelectric substrates by means of interdigital transducers (IDT). The most commonly used types of SAW sensors are also presented, as well as widely recognized characteristics of this type of gas sensors and their importance for sensor applications.

The application of surface acoustic waves (SAW) in gas sensors is an example of the interdisciplinarity of modern scientific and technical advances. In sensors of this type, phenomena from both acoustics, acoustoelectronics, and piezoelectronics, and physical chemistry of thin layers and layer structures are used. Traditional and modern microelectronic technologies and corresponding electronic circuits are also used.

In the field of acoustics, it is primarily the phenomenon of propagation of acoustic surface waves [15, 46], while in the field of acoustoelectronics and piezoelectronics, it is the way of excitation of these waves in piezoelectric crystals by means of interdigital transducers [63]. The phenomena of physisorption and chemisorption of gas and vapor molecules by various types of sensor layers and structures, produced in the area of SAW propagation, belong to the physics and chemistry of surface [1]. The production of interdigital transducers and the sensor layers and structures themselves, which can generally be called sensor elements, lies in the area of traditional photolithography and modern thin-film microelectronic technologies. The use of appropriate electronic systems to obtain measurable measurement signals and their processing is the last stage of work on the design of SAW sensors.

In each of the above-mentioned research areas, development can and should be attempted in order to obtain even better performance parameters of the designed sensors. However, it should be borne in mind that these ranges are interrelated. The physicochemical interactions of sensor layers and structures with vapor and gas molecules depend to a large extent on the technology of their implementation. On the other hand, the possibilities of influencing the phenomenon of SAW propagation depend, inter alia, on the geometric configuration of the sensor element. In this area,

W. Jakubik, *Application of Surface Acoustic Waves in Gas Sensors*, Springer Series in Surface Sciences 69, https://doi.org/10.1007/978-3-031-89971-3_2

the author's important contribution to the development of hydrogen sensors with SAW is noted through the introduction of bilayer sensor structures of phthalocyanine-palladium and WO_3-palladium, together with theoretical analysis of acoustoelectric interactions.

The general characteristics of SAW sensors are a consequence of all the above-mentioned interrelated ranges and stages of research, which will be developed to a greater or lesser extent in the later part of the chapter. This will make it possible to explain both the high efficiency of SAW sensors, in general, and the reasons for the high sensitivity of this type of gas sensors.

2.1 The Essence and Features of SAW Sensors

From the point of view of sensor applications, an important feature of SAW is the fact that there is a natural possibility to influence the conditions of its propagation. As a result of the action of various external factors (e.g. additional mechanical stresses T_{mz}, accelerations a_z, pressures p_z, temperatures T_z, electric fields E_z, or flows Q_z), these conditions can be altered and thus disturb the wave propagation in this way. If we can measure the resulting disturbances (by measuring the velocity, attenuation, or phase of the wave), then we get information about the size of the external perturbation factor (Fig. 2.1). However, special care shall be taken to ensure that during the observation of one of the wave's propagation perturbations, the influence of all other perturbations is reduced as much as possible by appropriate stabilization or eliminated by appropriate correction methods.

The relatively high ease of obtaining SAW propagation disturbances is due to the high energy density of this wave in the surface area. For the most commonly used SAW powers per beam width unit of 0.1–1 W/m, this density is on the order of 10^4–10^5 J/m^3. This is a simple consequence of locating mechanical displacements near the surface along which the wave is spreading. These displacements, depending on the type of wave, disappear at distances on average from one to several wavelengths from that surface.

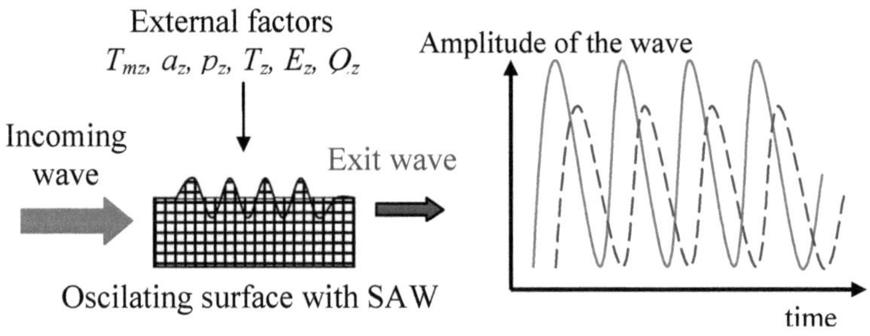

Fig. 2.1 The essence of the SAW sensor operation

When SAW propagates in a piezoelectric substrate, an electric field is associated with it, which disappears exponentially with the distance from the surface of the substrate. The disappearance of the wave amplitude is proportional to the factor $\exp(-k_0 y)$, where k_0 is the SAW wave number and y is the distance from the propagation surface. For SAW with a smaller wave number, i.e. longer wavelengths and lower frequencies, the electric potential associated with it will disappear more slowly. This feature opens the possibility of using larger ranges of thickness of sensor layers and structures characterized by measurable electrical conductivity, e.g. semiconductors. Changes in electrical conditions near the surface of the SAW propagation can affect basic parameters of its propagation, such as phase velocity or attenuation.

Another important feature of SAW is its relatively small speed of propagation compared to the speed of electromagnetic waves. Overall, it is about 10^5 times smaller, which is of great practical importance when determining its relative changes, caused by disturbance by various external factors. Possible relative changes, such as wave speed, are correspondingly larger.

The natural ease of access to the wave propagation area, which can be changed structurally; the high energy densities in the surface area; the associated electric field for piezoelectric substrates; and the small propagation velocity values are the main reasons for the potential applications of SAW in various types of sensors. In addition to vapor and gas sensors, waves of this type are used, among others, for the design of acceleration sensors [20], a whole group of liquid phase sensors, including biosensors [39], electric fields [21], as well as the flow sensors [4] and pressure sensors [7, 60].

2.1.1 The Essence of the Operation of SAW Gas Sensors

For gas sensors, in the SAW propagation area, a suitable sensor element has to be made, the material of which has to be matched to the properties and range of changes in gas concentration, as well as the operating temperature. This element in the form of a thin layer (with single, double, or even multiple structure) has to be placed in the SAW path because the larger the path of the wave interaction with the sensor element, the greater the changes in its propagation parameters can be. The basic configuration of the gas sensor with SAW and the thin-film sensor element is shown in Fig. 2.2a, while Fig. 2.2b shows a sensor with an excited surface wave, where the deformation scale has been greatly expanded for illustrative purposes. The thickness of the sensor element should be many times smaller than the length of the propagating surface wave ($h \ll \lambda$). This condition provides a weak perturbation of the wave propagation because it propagates at the surface of the substrate, without penetrating the sensor structure once it is manufactured. Too large thicknesses of the sensor layers (above 5% of the wavelength) result in high attenuation of the surface wave, making it much more difficult to excite and propagate. In addition, in such cases, it is possible to excite so-called Love waves, which propagate in layer systems with thicknesses comparable to the wavelength [6].

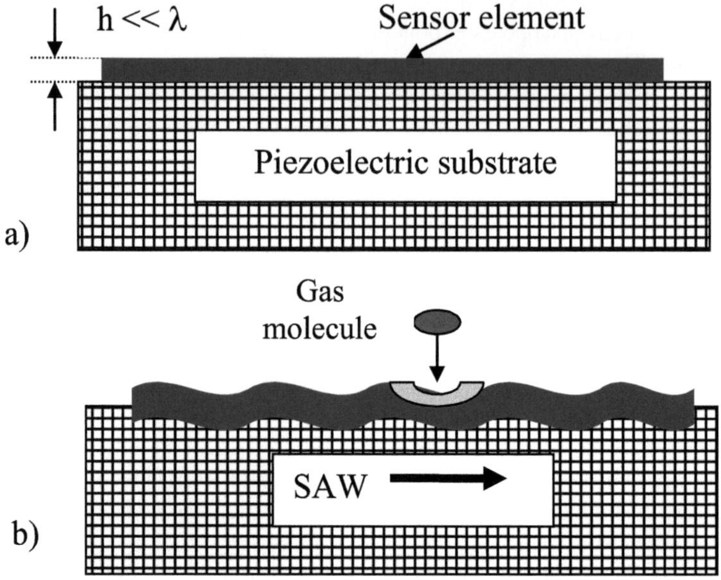

Fig. 2.2 Operation principle of SAW gas sensor with thin-film sensor element (**a**), and with excited SAW (**b**) with hypothetic adsorption place

The basic task of thin-film sensor elements is to change the propagation conditions as much as possible, as a result of physicochemical interactions with vapor or gas particles in their immediate environment. Depending on the type of materials used (dielectric, semiconductor, or metallic) and the geometry of the structure of the sensor elements produced (single, double, or multiple layers), their mechanical and/or electrical parameters change. This causes a disturbance of the wave propagation conditions, which results in changes in its speed and/or attenuation. The resulting relative changes in wave velocity are generally not large, they are of the order of magnitude 10^{-6}–10^{-3}. By measuring these very subtle changes, information about the concentration of the vapor or gas can be obtained. For this purpose, oscillator systems with a delay line are used (see Sect. 2.2.3), which allow precise monitoring of changes in the speed of the SAW by measuring their oscillation frequency. In some cases, specialized spectral analyzers are used, which additionally allow the determination of changes in the wave amplitude and phase under laboratory conditions [38].

The wave frequency f in the oscillator system should generally be represented as a complex (involved) function:

$$f \sim F\{v[p.s.e(c)]\} \tag{2.1}$$

where

v—velocity of SAW propagation, *p.s.e*—parameters of the sensor element (i.e. surface density, elastic constants, electrical conductivity), and *c*—concentration of vapor or gas in the immediate surroundings of the sensor element.

The function $p.s.e(c)$ represents a change in the parameters of the sensor element depending on the concentration of vapor or gas. This function is usually not known, but its effects are observed, usually in the form of changes in the velocity $v(p.s.e)$, which in the oscillator system is visible as a change in frequency. The most commonly used theoretical models of these relationships will be presented in Chap. 3.

SAW gas sensors, due to their specific characteristics, have a relatively low detection threshold, which is why they are ideal for detecting small concentrations of vapors or gases. We are looking for sensor elements that will change their parameters in the presence of gas molecules, so that this has a significant effect on SAW propagation.

The essence of the use of SAW in gas sensors is that even small changes in the parameters of the sensor layers or structures have a large effect on its propagation. The most common mass sensitivity results from the dependence of the speed and attenuation of SAW on the change in the mass of the sensor element as a result of the selective absorption of vapor or gas molecules from its environment and is defined as [59]

$$S_m = \lim_{\Delta m \to 0} \frac{1}{v_0} \frac{\Delta v}{\Delta m} \tag{2.2}$$

where

Δm—means change in the surface mass density of the sensor element (in g/cm^2),
v_0—is the SAW velocity on the free surface of the piezoelectric substrate, and
Δv—is the change of SAW velocity as a result of the change in mass density Δm.
The most commonly used unit of mass sensitivity is cm^2/g.

In the oscillator system (see Sect. 2.2.3), there is another way of defining mass sensitivity [59]:

$$S_m^f = f \cdot S_m \tag{2.3}$$

where S_m is defined by formula (2.2). The unit often used then is (Hz/ng)cm^2. Mass sensitivity in the oscillator system results from the multiplication of the sensitivity S_m (formula 2.2) and the frequency f of the excited acoustic wave. Negative mass sensitivity values indicate that the increase in mass causes a decrease in the speed of the wave propagation and the corresponding frequency in the oscillator circuit.

In addition to mass sensitivity, gas sensors with SAW can be characterized, already mentioned in the introduction, by acoustoelectric sensitivity. In this case, the sensor element must have an appropriate electrical conductivity, relative to the product of the SAW velocity and the electrical capacitance per unit length towards the wave propagation in the piezoelectric substrate. The definition of this type of sensitivity and its specific characteristics will be discussed in Chap. 3 (see Sect. 3.5.4.3).

In the case of gas sensors, the fact that dynamic changes of the sensor element continuously occur in the area of surface wave propagation, as a result of the corresponding component mechanical displacements over time, is of additional importance. The maximum amplitudes of these displacements, for medium acoustic power densities, are at the level of ~10^{-5} λ [19], which for a wavelength of $\lambda = 80$ μm gives a value of about 0.8 nm. Such small amplitudes do not cause mechanical damage to the sensor layers and sensor structures, what is more, they cause a certain kind of acoustic stimulation. The sensor element undergoes periodic deformation due to surface wave propagation increasing its absorption and desorption capabilities of gas molecules (Fig. 2.2b) This feature is characteristic only for SAW sensors and is an additional reason for their high sensitivity [22].

2.1.2 Physicochemical Phenomena in SAW Sensors

In gas sensors with SAW, there are phenomena of the physisorption and chemisorption of vapor or gas molecules on the surface, as well as absorption in the volume of the sensor elements produced in the wave propagation area. Which of the phenomena occurs or has a greater influence is determined by the properties of both the sensor element and the molecules interacting with it, which are in the gas phase. Among the first can be mentioned: the type of material, its manufacturing technology, and the resulting internal structure, i.e. the size of the grains in the case of polycrystalline samples, the morphology, the level of purity, and admixture. In the case of gas or vapor molecules, an important parameter is their electron affinity, which determines, among other things, whether electrons are given or received by them. The concentration of the gas and the temperature at which the gas and vapor interact with the sensor element are also important factors.

The influence of water vapor is also important. The occurrence of physisorption phenomena is more beneficial in terms of obtaining shorter response and regeneration times of the sensor. This is due to the energy of the bonds between the gas molecules and the sensor element, which in these cases are about 40 kJ/mol [44]. On the other hand, chemisorption, requiring higher binding energies of ~80–400 kJ/mol, allows obtaining greater changes in the parameters of the sensor elements, at the expense of longer response and regeneration times. In some situations, such strong interactions can result in structural changes in the material of the sensor element. This is the main cause of sensor instability, as its response will not be repeatable for the same concentrations and measurement conditions. The capabilities and nature of these phenomena determine the selectivity of the sensor.

If the sensor element is sensitive to the presence of only one type of gas or vapor molecules, then we can talk about excellent selectivity. In reality, however, this is never the case. Practically all currently known sensor materials intended for a certain gas also exhibit interactions with other gases, which leads to so-called *cross-sensitivity*. For the correct operation of a chemical sensor, it is important to sensitize the sensor element only to the presence of a given gas, i.e. to obtain very low cross-sensitivity. In justified cases, the type of material, the amount of additives, and the individual technological steps are the secrets of the companies producing a particular type of sensor. Most of the solutions are also subject to patent protection.

2.1.3 Features and Importance of SAW Gas Sensors

The characteristics of the SAW sensors result both from the properties of the surface waves themselves and from the properties of properly constructed sensor structures.

The widely recognized characteristics of these types of sensors are high sensitivity, compatibility with microelectronic technologies, satisfactory stability, fast real-time response, and relatively simple output signal processing (often in the form of frequencies). Among their utility characteristics are compactness, small dimensions, lightness, portability, and relatively low price. The ability to use much lower sensor element temperatures (~30–150 °C) compared to resistance sensors (~300–600 °C) is also very important. This leads to lower energy consumption during operation. Another important feature is the natural ability to create matrix circuits, by making several sensors with smaller dimensions on a common substrate [58].

The high sensitivity makes it possible to detect trace amounts, i.e. very small concentrations of vapors, gases, or volatile fractions of chemical compounds. Concentrations of certain air polluting gases such as NO_x, NH_3, SO_2, CO, and chemicals (causing paralyzing effects on the nervous system or human skin) can be detected at *ppm* (10^{-6}) and even *ppb* (10^{-9}) concentrations. This is of great importance for the protection of the environment, as well as human health and even life. Depending on the sensor element used, the high sensitivity of the SAW sensors can be used to monitor trace amounts of organic phosphorous compounds, aromatic hydrocarbons, saturated and chlorinated hydrocarbons, as well as ketones, alcohols, and water vapor [28].

Several modern technologies used in microelectronics can be identified in the manufacturing processes of SAW sensors: MIC (*microwave integrated circuits*), MEMS (*micro-electromechanical systems*), CMOS (*complementary metal-oxide-semiconductor*), and CCD (*charge coupled devices*) [28]. In addition, the basic process, especially for sensors with SAW, is photolithography. This technology is repeatable, and its bulkiness allows the price to be reduced to a few dollars for a single SAW filter (according to Siemens, for example). Methods such as CVD (*chemical vapor deposition*), PVD (*physical vapor deposition*), magnetron sputtering, and spin and spray coating are used to produce sensor elements in the form of thin or thick layers. It should be added that SAW modules work very well with classical electronic circuits, e.g. with feedback loop generators with appropriate electronic amplifiers.

The stability of SAW sensors is related to both the parameters of the sensor element used and to the parameters of the piezoelectric substrate itself. The possibility of using lower exposure temperatures affects the stability of the sensor elements used. On the other hand, the high sensitivity of some piezoelectric substrates (e.g. lithium niobate $LiNbO_3$) to temperature changes causes a deterioration in stability. A phenomenon called thermal drift is observed, i.e. a systematic slow decrease or increase of the sensor signal over time. The influence of small fluctuations in temperature, as well as external atmospheric pressure, can be limited by the use of the so-called differential system, i.e. two identical acoustic lines made on the same piezoelectric substrate (see Chap. 4). This partially compensates for the influence of temperature, increasing the stability of the entire sensor system.

It is noteworthy that substrates with a high temperature coefficient (less stable) also have a high electromechanical coupling coefficient. This is important for so-called acoustoelectric interactions, related to changes in surface electrical conductivity caused by the interaction of sensor layers with gas molecules.

The response times of SAW sensors depend on the sensor elements used, the concentration of the substance in the surrounding atmosphere that is being detected, and the operating temperature. These range from a few seconds to a few minutes. There is sometimes ambiguity in defining response times. Some authors accept 90–95% change in the frequency signal, while others lower it to 66%, which shortens the analysis time. Very often, only a time to reach is determined, e.g. 70% of the maximum change in the frequency signal, which shortens the overall analysis time.

In the case of SAW sensors, frequency is very often the output signal. It is widely known that frequency measurements are among the most accurate. The output signal in the form of frequency is easily processed in the further stages of the processing path (in the detection system), which is a very important feature of this type of sensor systems.

In addition to the above-mentioned important design features of SAW sensors, their utility features are also important. Compactness indicates that you can get the device as a single component forming a closed whole. In addition, it can be lightweight, with small dimensions, it is easy to move it to different places where it will be useful. The simple construction and standard electronic components will in turn make this type of sensors low price. As a result, it is possible to easily manufacture them depending on demand, even in large numbers.

Gas sensors with SAW, through their advantageous design and utility features, have great practical importance in areas such as protecting the environment (air pollution), protecting health, and even human life, e.g. against terrorist attacks using dangerous chemicals. There are also known military applications, such as detecting contamination by combat chemicals, detecting explosives, and even detecting landmines. The US space agency NASA uses sensors with SAW to monitor organic substances on board space shuttles and other space stations. The SAW sensors also enable wireless applications, which is important in hard-to-reach environments. In these cases, the SAW sensor system can be passive, meaning it does not require additional power, which is important for areas at risk of gas explosions.

2.2 Generation and Detection of SAW

SAW sensors are undoubtedly associated with piezoelectric materials, which enable the excitation of various types of acoustic waves. However, it has been many years since the discovery of the piezoelectric phenomenon itself [14], for its use in practice [36]. Historically, the first use of piezoelectric quartz as a sensor was the work of [53] on the study of ionizing radon radiation. It was the first to use the phrase *quartz electric balance* long before the use of quartz in quartz crystal microbalance (QCM) in 1964. Quartz was used to measure the charge accumulated in the capacitor as a result of ionizing radiation passing between its covers.

2.2.1 *Interdigital Transducer*

Currently, the most widespread method of SAW generation is the use of interdigital transducers (IDT) made on the surface of piezoelectric crystals [63].

With the discovery of this type of SAW generation, it became possible to use piezoelectricity in microelectronics. This resulted in many designs, especially band filters in the high frequency (HF) range, which were used for military purposes in radiolocation and telecommunications, and are now widely used in mobile telephony and satellite and cable television [20]. In addition, the development of HF band filters using interdigital transducers in radars and sonars can be observed [12]. Transducers of this type are also basic elements in sensors using different types of surface waves and piezoelectric substrates.

The most commonly used piezoelectric materials in SAW sensors are lithium niobate Y–Z (with electromechanical coupling coefficient $K^2 = 4.8\%$), quartz (0.11%), lithium tantalum (~1.6%), and more recently langazite (0.35%)—a new material that does not lose its piezoelectric properties even at high temperatures (~1200 °C) [10].

The simplest form of interdigital transducer is a periodic arrangement of metal electrodes and gaps between them, with equal widths of both electrodes and gaps, made on the surface of a piezoelectric crystal (Fig. 2.3).

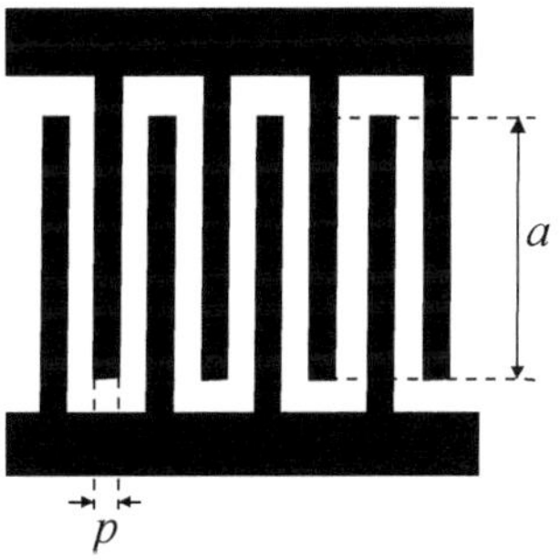

Fig. 2.3 The simplest interdigital transducer with electrode width equal to width of space between them p, and the electrode pair number $N = 4$, a—transducer aperture

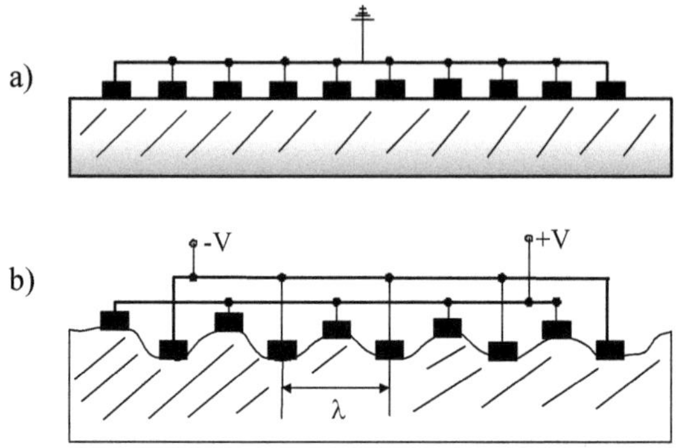

Fig. 2.4 Interdigital transducer in unexcited state (**a**), and after applying electric potential (**b**) generating SAW in both directions on piezoelectric crystal surface [56]

As a result of the piezoelectric phenomenon, when the appropriate electrical potentials are applied, mechanical displacement characteristics for surface waves are created (Fig. 2.4). The phenomenon of propagation of these displacements (in two directions from the transducer) is called an acoustic surface wave. The frequency of this wave is in the ultrasonic range.

The frequencies of the surface waves generated in this way depend both on the width of the interdigital transducers and the gaps between them, and on the wave speed, which is determined by the elastic properties of the piezoelectric substrate. The smaller the width of the electrodes and the gaps between them the shorter the wavelength λ and greater the frequency f of the wave, according to the formula:

$$\lambda = 4p, \quad f_0 = \frac{v_0}{4p} \tag{2.4}$$

where v_0—SAW velocity in the utilized piezoelectric substrate.

Current technological possibilities of traditional photolithography allow to make such electrodes from aluminum, sometimes from gold with widths in the range of 40–0.7 μm, which for the most typical piezoelectric substrates, such as quartz or lithium niobate, gives a wide range of possible frequencies 10 MHz–3 GHz.

More information on the properties and characteristics of interdigital transducers can be found in the review papers of Farnell [19], Śliwinski [54], Thomson and Stone [56] and Filipiak [20]. In the field of sensor applications, work on a more efficient but more complex surface wave excitation by *single phase unidirectional transducer* (SPUDT) can be seen [55, 61]. A suitable SPUDT system allows more acoustic power to be directed towards the receiver transducer, which greatly increases the efficiency of SAW generation.

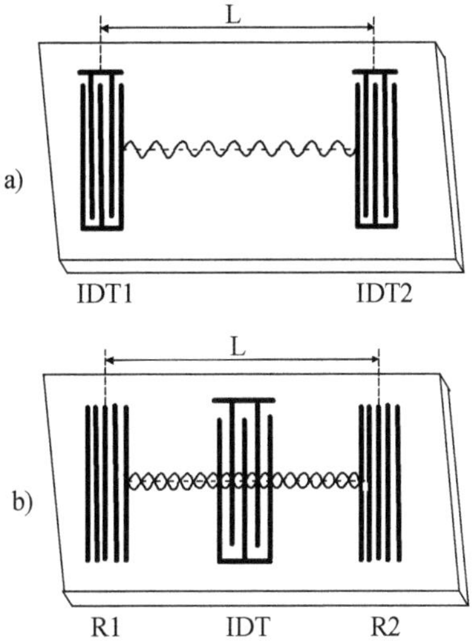

Fig. 2.5 Delay line (**a**) and one-port resonator (**b**) [9]

2.2.2 *Delay Line and Resonator*

SAW sensors come in two main types: with a delay line and as a resonator. The delay line system has two transducers: a transmitter (IDT1) and a receiver (IDT2), while the resonator is usually one in the middle of the structure (IDT) with additional metal strips on the sides (so-called *reflectors* R1 and R2) to reflect the wave back and produce a standing wave (Fig. 2.5). The edges of both ends of the piezoelectric substrate are sometimes beveled or covered with absorbent material, in order to attenuate the waves reflected in the main direction of propagation.

In the case of a delay line, the SAW excited in the transducer propagates over the surface of the piezoelectric and reaches the transducer with a certain time delay $\tau = L/v$, where L is the distance between the transducers and v is the phase velocity of the SAW. The electric field accompanying the wave induces, through an inverse piezoelectric phenomenon, the electrical voltages in the transducer. Since the typical lengths of the delay lines used are in the range of 5–20 mm, and the velocities depending on the substrate are 3000–3500 m/s, these delays are of the order of 1–7 μs. From the point of view of sensor applications, the use of a delay line is a much more convenient solution.

2.2.3 *Oscillator System*

Both the delay line and the resonator can be placed in the electronic oscillator system. Most often, the delay line is used as an element determining the frequency in the feedback loop of such an oscillator. The energy losses associated with the wave

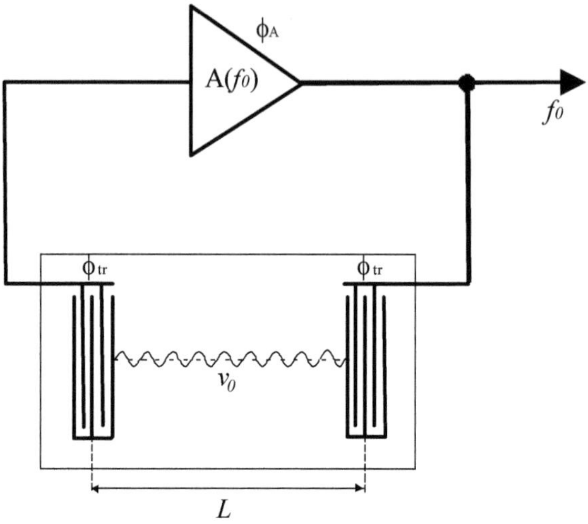

Fig. 2.6 Delay line in oscillator system

propagation are compensated and the signal with the characteristic frequency f_0 is excited and sustained (Fig. 2.6). The delay line together with the electronic amplifier represents a kind of resonator that uses a propagating surface wave with electrically compensated attenuation [9].

In electronic terms, the operation of the oscillator loop requires two conditions to be met:

$$\text{Amplification condition}: \quad A(f_0) \geq 1 \tag{2.5}$$

$$\text{Phase condition}: \quad 2\pi f_0 \tau_0 + \phi_A + 2\phi_{tr} = 2\pi n_0 \tag{2.6}$$

where

A—oscillator signal amplitude,

f_0—frequency of oscillation on the free surface of the substrate (without cover),

$\tau_0 = L/v_0$—signal delay on free surface,

L—the length of the acoustic track (between the transducers),

v_0—speed of SAW on the free surface of the crystal,

ϕ_A—phase shift in the amplifier,

ϕ_{tr}—phase shift introduced by a single interdigital transducer,

n_0—number of wavelengths between the transceiver and receiver transducers, $L = n_0\lambda$ (for $L = 18$ mm and $\lambda = 80$ μm, $n_0 = 225$).

Since the wavelength is determined by the constant geometry of the interdigital transducers, the frequency of the oscillation depends directly on the speed of SAW propagation in the delay line (formula 2.4). Increasing the speed of the SAW causes an increase in the oscillator frequency and vice versa, so the phase shift of the delay line for the excited oscillations (constant number n_0) does not change (component

$2\pi f_0\tau_0$ in formula 2.6). Hence the possibility of using the oscillator system in sensors with SAW, in which mainly the speed of the wave propagation changes as a result of the interaction of different types of sensor elements with vapor or gas particles.

The total phase shift caused by the delay line alone is several times greater ($2\pi f_0\tau_0$ ~ 80,000°) than the shift caused by the amplifier and transducers ($\phi_A + 2\phi_{tr}$ ~ 80°). Therefore, minor disturbances in the phase shift of the amplifier, caused by frequency changes due to the interaction of the sensor element with the gas, do not give rise to significant measurement uncertainties. The relative uncertainty in determining the frequency due to phase changes caused by a typical electronic amplifier (e.g. µA 733) is only 0.032% (based on catalog data), which gives a value of only 6.4 Hz at the maximum observed changes of ~20 kHz.

For piezoelectric substrates with high temperature coefficients, the velocity of SAW propagation is dependent on temperature changes. Therefore, depending on the substrate used, it is required to stabilize at least 0.4–0.01 °C, and in some applications even up to 0.001 °C. Thermal drift, caused by even a slight temperature gradient, is very easy to distinguish from typical interactions because it is very slowly variable over time.

2.3 Types of Sensors with SAW

Depending on the crystallographic orientation of the piezoelectric substrate, its thickness, the properties of the material center directly above the substrate, and the direction of placement of the transducer itself relative to the crystallographic axes of the crystal, different types of surface waves can be excited. They differ primarily in the directions of the excited mechanical displacements and in the energy density, i.e. its distribution in the area of wave propagation. Among the most famous ones that are used in sensors are Rayleigh waves (at the boundary of semi-infinite spring center—air), Lamb waves (formed in thin plates), and transverse-horizontal acoustic plate modes. Shear-Horizontal Acoustic Plate Mode (SH-APM), Love waves, and STW (surface transverse wave) [59]. STW or Love waves are sometimes classified as pseudosurface waves because they are not strictly surface waves [35].

The use of a particular type of acoustic wave in a given sensor type mainly depends on the directions and values of the excited mechanical displacements on the surface of the substrate, the density of the wave energy in the surface area, and the strength and complexity of the overall sensor design. As previously noted, the sensitivity of the SAW sensor strongly depends on the density of wave energy at the contact site, while the directions of mechanical displacements determine the possibility of application in a gas or liquid environment.

2.3.1 Rayleigh-Type SAW Sensors

In gas sensors, Rayleigh-type surface acoustic waves are most commonly used, abbreviated in English like SAW (surface acoustic wave). A wave of this kind at the boundary of two semi-infinite isotropic materials was theoretically predicted in Rayleigh's classic work [46]. Speaking of acoustic surface waves, this type of wave is most commonly considered [62].

Rayleigh's SAW is characterized by mechanical displacements in the perpendicular direction as well as parallel to the surface of the propagation. In addition, the composition of the upright is generally larger than the parallel one, and both lie in a plane perpendicular to the surface and parallel to the direction of the propagation (*sagittal plane*). The particle on the surface of the material moves along an elliptical path in a plane consistent with the direction of the wave propagation and perpendicular to the ground surface. In general, the propagation of SAW across the free surface of the crystal towards the z-axis results in displacements $\boldsymbol{u}$ of particles around their equilibrium positions, which change with depth y to the substrate as [41]

$$\vec{u}(x, y, z, t) = e^{(i\omega_0 t - k_0 z)} \sum_{j=x}^{z} u_j(y) e^{i\phi_j} \vec{e}_j \tag{2.7}$$

where $\omega_0 = 2\pi f_0$ is the circular frequency (pulsation), k_0 is the common constant of propagation, $u_j(y)$ is the component of displacements in the directions $j = x$, y, z, dependent on the depth of penetration, ϕ_j is the relative phase of these displacements in relation to u_z, and e_j is the unit vector. The component displacements $u_j(y)$ disappear exponentially with distance to the surface, so that a significant part of the wave energy is contained in the area of one to two wavelengths. The displacements thus disappearing are the cause, already mentioned, of the high energy density in the surface propagation area of SAW.

The Rayleigh wave is extremely sensitive to changes in the mechanical boundary conditions of the propagation. In the sensor system, this leads to the so-called mass effect, which is related to the mechanical parameters of the sensor element produced on the surface of the substrate in the area of the wave propagation area. In addition, in the case of piezoelectric substrates with an acoustic surface wave, an electric field is associated. The wave propagation in this case will also depend on changes in the electrical boundary conditions, which leads to the so-called acoustoelectric effect, i.e. the dependence of the wave speed and attenuation on changes in the electric surface conductivity of the layer or the entire sensor structure.

The ability to use two different sensor mechanisms at the same time is an important and specific feature of SAW gas sensors.

The decisive factor in which of the interaction mechanisms will be dominant is the type of sensor structure produced in the SAW propagation path. For non-conductive structures or structures with very low electrical conductivity (e.g. polyaniline layers), mass interactions will be important, i.e. the change in the mass of the structure will determine the accidental change in the wave velocity. Using such a mechanism, the

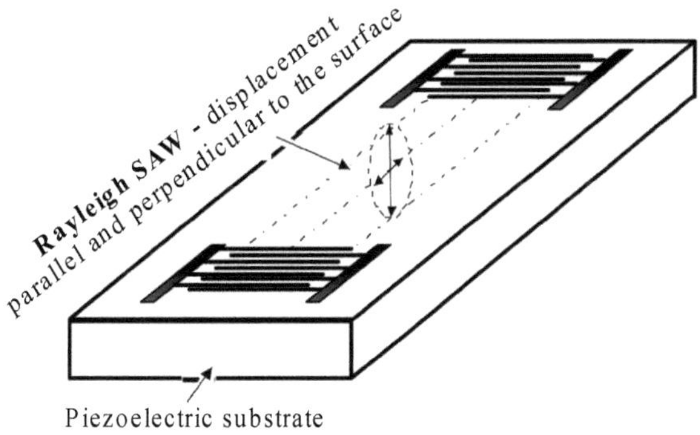

Fig. 2.7 Mechanical displacements in Rayleigh-type SAW excited on piezoelectric substrate in single delay line configuration

first SAW sensor proposed by Wohltjen [65] worked. It used a quartz substrate of type ST, which together with interdigital transducers enabled the creation of Rayleigh-type SAW (Fig. 2.7). Surface wave of this kind was historically used for the first time in a single-line system to analyze the gas phase of organic compounds using thin layers of non-conductive polymers [65]. Theoretically, the mass detection limits presented in this work are up to 3×10^{-15} g. On the other hand, practically confirmed typical detection limits for mass sensitivity, e.g. for SAW system with a frequency of 100 MHz, are values of about 250 pg/cm^2 [59].

For sensor structures with adequate electrical conductivity, in addition to mass detection, an additional acoustoelectric effect may occur, which in some cases leads to many times greater sensitivity of the gas sensor [48, 57]. Therefore, gas sensors with SAW on a piezoelectric substrate, for such structures, are characterized by acoustoelectric (AE) sensitivity. One of the parameters determining the AE sensitivity is the electromechanical coupling coefficient of the piezoelectric material K^2, which is expressed as the ratio of energy stored in the electric field to mechanical energy. However, it is crucial to match the electrical conductivity of the layer, or the entire sensor structure, to the so-called area of strong acoustoelectric interactions, which will be discussed in detail in Chap. 3. The most commonly used piezoelectric substrates, together with their relevant parameters, for Rayleigh's SAW sensors are presented in Table 2.1.

The stability of the Rayleigh SAW sensor system is dependent on the temperature coefficient of the piezoelectric substrate used. Even small temperature fluctuations (depending on the substrate ~0.01–0.1 °C) can cause significant changes in the speed of surface wave propagation. This effect can be reduced by using less sensitive piezoelectric substrates, such as quartz ST. However, it has a much lower electromechanical coupling factor. This type of substrate reduces the possibility of the occurrence of acoustoelectric effects.

In addition to the most basic interactions, mass and acoustoelectric, for certain types of polymer layers, a coupling can occur involving a change in the speed of the SAW as a result of a change in their viscosity [43]. This type of interaction is relatively rarely observed because it is a consequence of changes in the viscosity parameters of the sensor elements used.

Table 2.1 Piezoelectric substrates enabling Rayleigh SAW excitation

Substrate	Velocity, m/s	K^2, %	TCF, ppm/°C	C_s, pF/cm
Quartz ST	3158	0.11	0	0.5
$LiNbO_3$ YZ	3488	4.8	94	4.5
$LiTaO_3$ 128° YX	3992	5.3	75	5.0
$LiTaO_3$ 77° YZ	3254	0.72	35	4.5
Langasite[a] $La_3Ga_5SiO_{14}$	2740	0.35	25	5.0

TCF—temperature coefficient of frequency, C_s—capacity per unit length of the interdigital transducer pair [35]
[a] (FOMOS-MATERIALS, 16 Buzheninova Street Moscow, 107023, Russia)

Rayleigh-type SAW sensors, due to the generated displacements on the piezoelectric surface, are primarily intended for gas phase detection. In some cases, their high sensitivity can also be used to analyze the properties of liquid helium (^{4}He) in the low temperature range [3, 4].

For gas sensors, different types of sensor elements are used, which have to be produced in a thin-film form, i.e. their total thickness is many times smaller than the wavelength of the propagating wave. Meeting this condition ($h \ll \lambda$) in systems with a sufficiently long wavelength (SAW wavelength ~50–100 μm) opens up some possibilities for new sensor structure configurations.

Double- and even triple-layer structures with thicknesses of up to hundreds of nanometers still meet this requirement. This is a completely novel solution that has not been used before in this type of sensor systems with SAW (Fig. 2.8). However, the total thickness of the structure must be much less than the wavelength of the propagating surface wave.

Previous attempts to use single palladium metal layers in Rayleigh's SAW sensors for hydrogen detection have not yielded satisfactory results [18]. A closer analysis of this problem has shown beyond any doubt that placing single metal layers on the surface of piezoelectric with SAW allows the use of only mass detection (a change in the layer's mass or possibly its elastic parameters). The SAW electric field that causes the acoustoelectric interactions is reduced to zero in this case. Only moving the metal

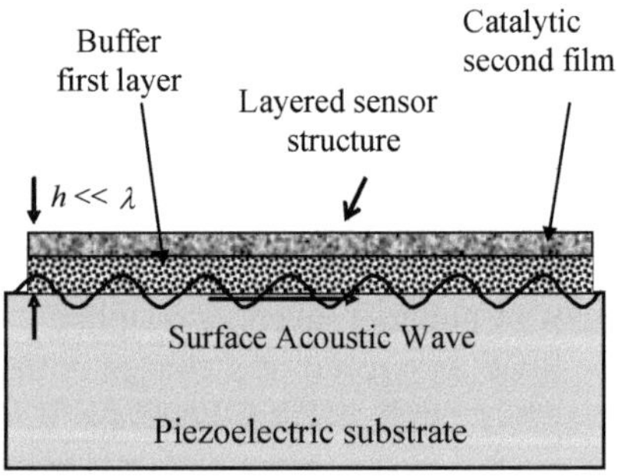

Fig. 2.8 Idea of layered structures, the overall thickness $h \ll \lambda$

layer above the crystal surface at a distance of hundreds of nanometers ensures the use of strong acoustoelectric interactions, which can be many times greater than mass. In many cases, acoustoelectric interactions have been confirmed for this type of layered structures [32, 31].

The displacement of the metallic layer above the surface of the crystal is carried out in the layered structures, by producing the corresponding buffer layers. The role of the buffer layer (the first on the surface of the crystal) is not limited to a geometric function only. It can additionally absorb the gas molecules that have been adsorbed by the catalytic layer, and can further increase its active surface due to the so-called *surface expansion.*

The development of this group of gas sensors involves the search for new sensor materials or a new configuration of the entire sensor structure. Among the new materials are polymers, e.g. polypyrol [2], polyanilines [49], as well as carbon nanotubes (CNT) [52] and two-layer structures like polyaniline-WO_3 [50].

2.3.2 *Other Types of Acoustic Sensors*

The most common, other types of acoustic sensors are shown in Fig. 2.9. They will be described in brief [59].

SH-APM (*Shear-Horizontal Acoustic Plate Mode*) is an acoustic plate mode with horizontal polarization, proposed by Ricco and Martin [47] in a sensor for working in the liquid phase. The wave energy is not radiated to the liquid, resulting in parallel displacements on the surface of the substrate (Fig. 2.9a). Waves of this

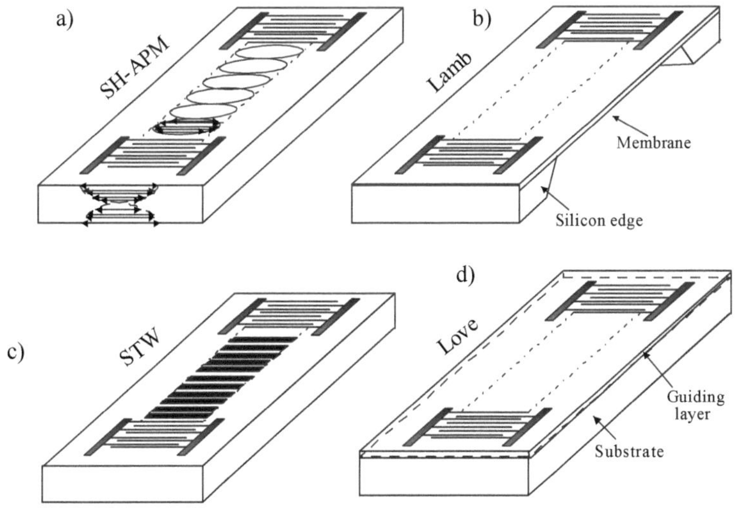

Fig. 2.9 Other types of acoustic sensors: SH-APM (**a**), Lamb (**b**), STW (**c**), and Love (**d**)

type are possible to excite in the ST quartz in the direction of X-propagation. The disadvantage is that the wave energy is distributed over the entire volume of the substrate, which reduces the sensitivity. Therefore, it is not strictly a surface wave, which is why this type of sensors can be classified as volumetric systems that excite surface by means of interdigital transducers [13]. The displacements on the surface are parallel to it, whereas in volume we obtain the superposition of the wave falling and reflected off the bottom surface of the substrate. Only by reducing the thickness of the substrate can satisfactory sensitivity be achieved, at the expense of its fragility. On the other hand, it is advantageous to use both surfaces of piezoelectric substrates, since plate modes (APMs) are reflected off each of them. The mass sensitivities obtained are several times smaller (six-, sevenfold) than for Rayleigh wave sensors of the same frequency. This type of acoustic sensors is used in liquid phase viscosity determination and in biosensors [42].

The sensor system using **Lamb** plate waves [64], most commonly of the antisymmetric type (then marked as FPW—Flexural Plate Wave), is basically designed for the detection of the liquid phase (Fig. 2.9b). The wave propagation occurs in a thin membrane produced by etching in silicon. The antisymmetric mode of Lamb wave in very thin membranes (with a thickness of 2–10 μm, which is 5% or less compared to the wavelength) has a lower phase velocity than the velocity of the sound wave in the liquid waveguide. Therefore, the thin membrane functions like a waveguide, without losing the energy radiated to the liquid, despite the perpendicular displacements on the surface. Lamb waves can be thought of as the propagation of two Rayleigh waves on each surface of the thin membrane. Due to the low phase velocity of the wave, the frequencies of this type of sensor are also low (5–20 MHz) compared to other acoustic wave systems. Due to the small thickness of the membranes where the Lamb waves propagate, the obtained mass sensitivities are very large, reaching absolute values of about 500 cm^2/g, which at a frequency of 10 MHz gives a value of 5 $(Hz/ng)cm^2$. A unique feature of Lamb's wave sensors is their unique sensitivity to changes in liquid density. Possible resolution for this density (at 10 MHz) is even 1 g/m^3 [59].

The sensor with a surface transverse wave **STW** (surface transverse wave) was first presented by Baer et al. [5]. Thin metal strips placed in the path of the wave reduce its speed and act as a waveguide as a result. The wave energy is thus more concentrated in the surface area, increasing the sensitivity to disturbances at the surface. The structure of this type of sensor is shown in Fig. 2.9c. For this type of wave, quartz ST can be used, where the X-axis is parallel to the metal strips. STW waves are produced by the generation of periodic surface disturbances, e.g. in the form of metal strips for SSBW waves (*surface skimming bulk wave*). STW waves are polarized like SH-APM, i.e. horizontally, but show higher mass sensitivities as a result of the surface energy concentration. The obtained absolute measurable values of surface density changes are 180 g/cm^2, with a detection limit ~0.2 ng/cm^2 for the detection of human immunoglobulin type G (HIgG) [5]. STW waves in the resonator system were also used to detect moisture through a new material labeled as PPAa (plasma polymerized allylamine) [67].

A sensor using **Love** waves is used for liquid phase detection and has historically been presented independently by two groups Gizeli et al. [23], and Kovacs and Venema [37]. A thin waveguide layer is produced on a piezoelectric substrate with interdigital transducers manufactured. The transducers are located in the area between the waveguide layer and the substrate (Fig. 2.9d). If the circuit is well designed, most of the acoustic energy is located in the waveguide. For a SiO_2 waveguide on a quartz substrate (ST section with propagation direction perpendicular to the X-axis), its optimal thickness is about 18% of the propagating wavelength. Possible applications of this type of sensor are biosensors, e.g. for the liquid phase, where the active layer is placed on the surface of the waveguide. In other solutions, the waveguide layer itself can function as the active layer [24]. The mass sensitivity of Love wave sensors depends on the frequency used, the waveguide material, and its thickness normalized to the wavelength. The absolute sensitivity values obtained are 450 cm^2/g and for 200 MHz: 90 $(Hz/ng)cm^2$.

An extremely valuable feature of sensors with Love wave is the ability to work in a liquid medium. For this reason, it is possible to design biosensors to detect different types of microorganisms, including bacteria [25, 27]. Sensor elements in biosensors called receptors can be of different types, often they are enzymes or antibodies. In addition, immobilization is used to enable proper absorption [45].

2.3.2.1 Types and Stages of Acoustic Sensor's Development

At the end of this chapter, types (Fig. 2.10) and selected stages of development of the most commonly used acoustic sensors are presented (Table 2.2). They cover the major groups of these sensors and the most important work and events related to them, ranging from the discovery of the piezoelectric phenomenon to the use of bilayer and light-activated sensor structures.

2.4 Summary and Comparison of Acoustic Sensors

The chapter presents the general characteristics of SAW gas sensors. The most commonly used types of acoustic sensors are also discussed, giving their basic characteristics and application possibilities. Table 2.3 presents their comparison by taking into account sensitivity, noise level, performance, mechanical strength, complexity of execution, applications, and level of development.

Mass sensitivity is closely related to the energy density near the substrate surface after which the wave spreads. Due to their small membrane thicknesses (2–10 μm), Lamb wave systems are characterized by high mass sensitivity, at the expense of greater complexity of execution compared to wafer-based crystalline substrates.

Love wave systems are also characterized by fairly high mass sensitivity. Their execution is only slightly more complex than Rayleigh wave, APM, or STW systems. In the case of Love waves, an additional waveguide layer is made on the surface of the crystal along with the transducers, while for STW waves a metallic layer is needed on the surface of a piezoelectric.

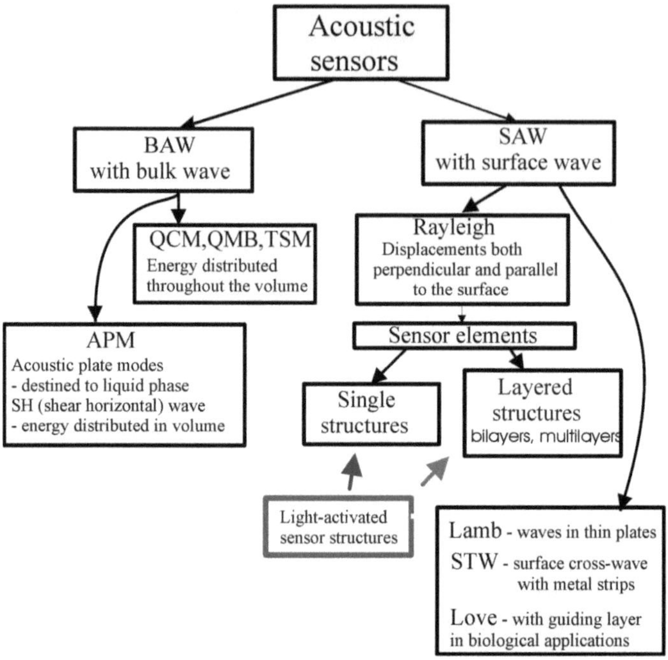

Fig. 2.10 Major acoustic sensor groups with Rayleigh's SAW sensor and with layered structures

APMs have lower mass sensitivities compared to other types of sensors, but their advantage is the possibility of using two piezoelectric plate surfaces, which is particularly important in liquid phase detection.

Rayleigh-type SAWs cannot be used in a liquid environment because they radiate a significant amount of wave energy into the liquid, resulting in high attenuation.

The sensitivity of a particular sensor system is related to noises and frequency instability of the oscillator system. Higher sensitivity along with higher noises will not make for a better sensor. Typically, the sensitivity increases with increasing frequency, but just as often this causes an increase in noise levels and the resulting gain is small. In addition, for higher frequencies, there are problems with electrical connections, as their impedance between the electronic system and a sensor begins to play a role.

A big advantage of all the discussed types of acoustic sensors is their design technology, which, in addition to BAW and Lamb systems, uses planar methods. A thin metal layer (vacuum evaporation) is produced on the piezoelectric substrate, followed by photolithography by transducers and other elements (e.g. metal strips for STW). In the case of Love wave systems, an additional process is needed to produce the waveguide (e.g. SiO_2 on the ST quartz surface). For Lamb waves, the technology is more complex, as it is required to digest in silicon to obtain sufficiently thin membranes and additionally to deposit piezoelectric layers (e.g. ZnO), because silicone is not piezoelectric.

Table 2.2 The most important stages and works related to the development of acoustic sensors using piezoelectric phenomenon

• The discovery of the piezoelectric phenomenon [14]
• The first use of piezoelectric material (quartz) as a sensor—change in capacitor capacitance as a result of passage of ionizing radiation ("quartz electric balance") [53]
• Bulk Acoustic Wave Resonator [51]
• The first practical application of quartz crystal microbalance (QCM)—frequency change due to mass change [36]
• IDT (interdigital transducers)—excitation of surface waves in piezoelectric crystals—combination of piezoelectricity with microelectronics—filters, radars, sonar, telecommunications (filters, demodulators, oscillators) [63]
• SAW device—the first sensor system with Rayleigh type SAW using IDT (polymer layers for VOC—volatile organic solvents)—greater sensitivity, because higher frequencies are used compared to Bulk Acoustic Wave (BAW) [65]
• Sensor system type SH-APM enabling work in the liquid phase, quartz type ST propagation direction X—the wave is not radiated to the liquid, because the displacements on the surface are parallel to it (shear horizontal) [47]
• Sensor system with Lamb wave (type—flexural plate wave)—wave velocity less than the velocity in the liquid phase—no energy radiation despite perpendicular displacements to the surface of the waveguide [64]
• STW (surface transverse wave) sensor system with thin metal strips on the path of wave propagation, which reduces its speed—the system acts like a waveguide, increasing the concentration of energy at the surface [5]
• Love's wave system for liquid phase detection—most of the acoustic energy is located in an additional waveguide layer on the surface with IDT transducers [23] and [37]
• SAW sensor systems with phthalocyanine-palladium bilayer structures for detection of hydrogen in the pre-explosion range of concentrations in the air [33]
• SAW sensor systems with light-activated photoconductive polymer films for dimethyl-methylophosphor (DMMP) detection—simulant of chemical warfare agents (CWAs) [34]

Possible applications include the use of quartz crystal microbalance (QCM) systems to measure the thickness of thin layers (so-called monitors or quartz resonators) and as extremely precise hygrometers, with a detection level of up to 40 ppb, developed for the analysis of natural gas [26]. In addition, QCMs are used as chemical dosimeters—especially mercury content analyzers [40].

Sensors with surface waves find major applications in the detection of various types of organic compounds and gases, as well as for determining the content of various substances in the liquid phase. An example here is the detector S-sens K5 by the German company Nanofilm. It is a five-channel sensor with SAW for the detection of various types of biological substances, such as platelets in human blood (thrombocytes), and even certain types of cancer cells. The most important elements of this device are the sensor structures, which can be thin layers of gold and SiO_2 [25].

Table 2.3 Comparing the properties of acoustic sensors

Wave type/frequency range in MHz	Mass (S_m) or acoustoelectric (S_{AE}) sensitivity/noise level sensitivity description/typical values	Work environment	Durability/complexity of work	Application	Development level
QCM/3–15 bulk resonator	Medium/low $S_m = -2/\rho_p\lambda$ 10–20 cm^2/g	Gas, liquid	Low/low	Gas detection, thickness measurements	Large-scale production
Rayleigh/40–600	High/low $S_m = -K(v_P)/\rho_p\lambda$ 100–200 cm^2/g	Gas	High/low transducers on the crystal surface	Gas detection, pressure measurements	Low-scale production
Rayleigh + layered structures/40–50	Very high/low $S_{AE} = -3\sqrt{3}K^2/16v_0C_s$ ~ (2–9) × 10^3 S^{-1}	Gas	High/medium transducers on the crystal surface + layered structure	Gas detection	Laboratory prototypes[a]
Rayleigh + light-activated sensor structures/~200	Extremely increased/low S_{AE} ~ 200–300 times increased with matching activation to the layer	Gas	Medium (film degradation observed)/medium (light activation system)	Gas detection (DMMP simulant of sarin)	Laboratory prototypes[b]
APM/~100	Medium/low $S_m = -1/\rho_p d$ 20–40 cm^2/g	Gas, liquid	Medium/low transducers on the crystal surface	Gas detection, biochemical, viscosity	Laboratory prototypes
Lamb/~10	High/medium $S_m = -1/\rho_p d$ 200–1000 cm^2/g	Gas, liquid	Low/high silicon etching	Gas detection, biochemical density	Laboratory prototypes

(continued)

Table 2.3 (continued)

Wave type/frequency range in MHz	Mass (S_m) or acoustoelectric (S_{AE}) sensitivity/noise level sensitivity description/typical values	Work environment	Durability/complexity of work	Application	Development level
STW/40–100	High/low $S_m = -1/\rho_p b$ 100–200 cm^2/g	Gas, liquid	High/low metal strips on crystal surface	Gas detection, biochemical	Laboratory prototypes
Love/100–200	High/low $S_m = -1/\rho_p h_f$ 150–500 cm^2/g	Gas, liquid	High/medium waveguide on crystal	Gas detection, biochemical, viscosity	Laboratory prototypes

ρ_P—substrate density, $K(\nu_P)$—parameter dependent on Poisson coefficient, typically $1 < K(\nu_P) < 2$, λ—wavelength, K^2—coefficient of electromechanical coupling of substrate, v_0—speed of SAW on free substrate, C_s—sum of electrical permeability of substrate and environment, d—thickness of substrate, b—effective waveguide thickness, h_f—thickness of waveguide layer [17, 59]

[a] [30]

[b] [34, 66]

The company founded by Wohltjen Microsensor Systems Inc. Springfield USA, which was the first to use SAW systems for chemical analysis, manufacture's sensor chips with SAW on a mass scale. The SAW MiniCAD mk II, an American company, is also available for the detection of chemical warfare agents. The total analysis time of this device is about 1 min [11]. The response times, obtained by the author, for the detection of hydrogen in the air using prototype bilayer structures, reach values from a few to several seconds [29].

SAW systems are also used in the automotive industry to directly monitor the pressure in automotive wheels. Commercially available solutions of this type can be ordered as optional equipment priced at $50–$700. An important feature is the possibility of wireless applications, which is also used in humidity sensors [16].

Recent trends include the use of ever higher frequencies to increase mass sensitivities, e.g. working on higher harmonics (Kawalec and Pasternak 2008), and inductive and capacitive coupling of SAW transducers with electronics at high frequencies [8]. New sensor materials are also important, i.e. carbon nanotubes (CNT), palladium nanowires, and more efficient excitation of surface waves using *single phase unidirectional transducer* (SPUDT) [55, 61].

The author's contribution to the development of this field is the proposal to use sensor layer structures, e.g. semiconductor–metal type, which enables to make sensor systems with SAW with much greater sensitivity than for single layers. The research shows that the use of bilayer structures, e.g. phthalocyanine-palladium and metal oxide-palladium, to detect hydrogen in the air in the pre-explosion concentration range significantly increases the sensitivity of SAW systems. In this sense, the use of layer structures in Rayleigh-type SAW sensors constitutes a new group, which is shown in the diagram (see Fig. 2.10), representing the main types of acoustic sensors. The layered structures developed and studied by the author are an extension of the classic group of Rayleigh wave sensors with displacements both perpendicular and parallel to the piezoelectric surface. Therefore, they are mainly intended for detection in the gas phase. Just like single layers, layered structures must have a total thickness many times smaller than the propagating Rayleigh wavelength. Failure to meet this condition results in wave propagation not only in the surface area of the substrate, but also in the layer structure, which is characteristic of Love wave sensors.

One of the more recent trends developed by the author of the monograph and collaborators is the use of light activations for thin layers of photoconductive polymers in the SAW system. Sensor structures activated in this way show an extremely high sensitivity to detected substances, e.g. the regioregular poly-3-hexylothiophene (rrP3HT) photoconductive polymer has been used in the detection of trace amounts of dimethylomethylophosphor (DMMP) (simulant of poisons including sarin). The observed significant increase in frequency shifts can be interpreted as acoustoelectric interactions, i.e. shifting of the working point of the structure to a high sensitivity zone as a result of the light activation process. In the case of constant or slowly changing humidity, such designs are an easy way to develop fast and low-temperature gas sensors that can be compared to existing devices (see Chap. 6).

References

1. Adamson AW (1982) Physical chemistry of surfaces, 4th edn. Wiley, New York
2. Al-Mashat L, Tran H, Włodarski W, Kaner R, Kalantar-Zadeh K (2008) Polypyrrole nanofiber surface acoustic wave gas sensors. Sens Actuators B 134:826–831
3. Aoki Y, Wada Y, Sekimoto Y, Yamaguchi W, Ogino A, Saitoh M, Nomura R, Okuda Y (2004) Application of surface acoustic wave sensors for liquid Helium-4 and Helium-3. J Low Temp Phys 134(3/4):945–957
4. Aoki Y, Yamaguchi W, Wada Y, Sekimoto Y, Nomura R, Okuda Y (2003) Acoustic properties of liquid ^{4}He measured by Rayleigh-SAW. Phys B 329–333:234–235
5. Baer RL, Flory CA, Tom-May M, Solomon DS (1992) STW chemical sensors. In: IEEE ultrasonic symposium, vol 1. Tucson, Arizona, pp 293–298
6. Ballantine DS, White RM, Martin SJ, Ricco AJ, Frye GC, Zellers ET, Wohltjen H (1997) Acoustic wave sensors: theory, design and physico-chemical applications. Academic Press, New York
7. Baron T, Hermelin D, Romand J-P, Alzuaga S, Queste S, Rauch J-Y, Leguen J-F, Belgasem B, Ballandras S (2010) SAW pressure sensor on quartz membrane lapping. 10eme Congres Francais d'Acoustique, Lyon
8. Bender F, Lange K, Voight A, Rapp M (2004) Improvement of surface acoustic wave gas and biosensor response characteristics using a capacitive coupling technique. Anal Chem 76:3837–3840
9. Benes E, Groeschl M, Burger M, Schmid M (1995) Sensors based on piezoelectric resonators. Sens Actuators A 48:1–21
10. Berkenpas E, Kenny T, Millard P, da Cunha PM (2005) A langasite SH SAW 0157:H7 E. coli sensor. In: Ultrasonic symposium, vol 1. Rotterdam, pp 54–57
11. Bulletin Inf. MSA (2006) SAW MiniCAD mk II chemical warfare agent detector. 10074824, USA
12. Campbell CK (1998) Surface acoustic wave devices for mobile and wireless communications, vol 1. Academic Press, USA
13. Cheeke JDN, Wang Z (1999) Acoustic wave gas sensors. Sens Actuators B 59:146–153
14. Curie J, Curie P (1880) Développement, par pression, de l'électricité polaire dans les cristaux hémièdres à faces inclinées. C R Acad Sci Paris 91:294–295
15. Datta S (1985) Surface acoustic wave devices. Prentice-Hall, Enclewood Cliffs, NJ, ch 7, pp 179–186
16. Dierkes M, Hilleringmann U (2003) Telemetric surface acoustic wave sensor for humidity. Adv Radio Sci 1:131–133
17. Dorozhkin LM, Rozanov IA (2001) Acoustic wave chemical sensors for gases. J Anal Chem+ 56(5):399–416
18. D'Amico A, Palma A, Verona E (1982) Hydrogen sensor using a palladium coated surface acoustic wale delay-line. In: IEEE ultrasonic symposium, vol 1, pp 308–311. San Diego, 27–29 Oct 1982
19. Farnell G (1977) Surface wave filters elastic. In: Matthews H (ed) Surface waves. Wiley, New York, ch 1, pp 1–53
20. Filipiak J (2006) Czujniki przyspieszenia z akustyczną falą powierzchniową (Acceleration sensors with surface acoustic wave). Monografia Politechniki Częstochowskiej nr 121
21. Fransen A, Lubking GW, Vellekoop MJ (1997) High-resolution high-voltage sensor based on SAW. Sens Actuators A 60:49–53
22. Giriuniene R, Garska E (2000) Acoustoelectric gas sensor with casserite film. ULTRAGARSAS (Ultrasound) 35(2):27–29
23. Gizeli E, Goddard CR, Lowe AC, Stevenson A (1992) A love plate biosensor utilizing a polymer laser. Sens Actuators B 6:131
24. Gizeli E, Lowe Ch, Liley M, Vogel H (1996) Detection of supported lipid layers with the acoustic Love waveguide device: application to biosensors. Sens Actuators B 34:295–300

25. Gronewold TMA, Schlecht U, Quandt E (2007) Analysis of proteolytic degradation of a crude protein mixture using a surface acoustic wave sensor. Biosens Bioelectron 22:2360–2365
26. Guilbaut GG (1984) Applications of piezoelectric quartz crystal microbalances, vol 7. In: Lu, Czanderna (eds) Elsevier, New York
27. Howe E, Harding G (2000) A comparison of protocols for the optimisation of detection of bacteria using a surface acoustic wave (SAW) biosensor. Biosens Bioelectron 15:641–649
28. Hribsek M, Tosic D (2010) Analysis and modelling of Surface Acoustic Wave chemical vapour sensors. Acoust Waves 359
29. Jakubik WP (2005) Hydrogen detection by single and bilayer sensor structures in Surface Acoustic Wave system. J Phys IV 129:117–120
30. Jakubik W (2011) Surface acoustic wave-based gas sensors. Thin Solid Films 520(3):986–993
31. Jakubik W, Krzywiecki M, Maciak E, Urbańczyk M (2012) Bi-layer nanostructures of CuPc and Pd for resistance-type and SAW-type hydrogen gas sensors. Sens Actuators B. https://doi.org/10.1016/j.snb.2012.10.056
32. Jakubik W, Urbańczyk M (2001) Surface Acoustic Wave hydrogen sensor with multilayer structure-preliminary results. Proc SPIE 4514:194–200
33. Jakubik WP, Urbańczyk MW, Kochowski S, Bodzenta J (2002) Bilayer structure for hydrogen detection in a surface acoustic wave sensor system. Sens Actuators B 82(2–3):265–271
34. Jakubik W, Wrotniak J, Kaźmierczak-Bałata A, Stolarczyk A, Powroźnik P (2023) Light-activated SAW sensor structures with photoconductive polymer films for DMMP detection. Sens Actuators B 384:133597. https://doi.org/10.1016/j.snb.2023.133597
35. Kalantar-Zadeh K, Powell DA, Włodarski W, Ippolito S, Galatsis K (2003) Comparison of layered based SAW sensors. Sens Actuators B 91:303–308
36. King WH (1964) Piezoelectric sorption detector. Anal Chem 36:1735–1739
37. Kovacs G, Venema A (1992) Theoretical comparison of sensitivities of acoustic shear wave modes for (bio)chemical sensing in liquids. Appl Phys Lett 61(6):639
38. Kozlovsky NY, Malocha DC, Weeks AR (2011) A 915 MHz saw sensor correlator system. IEEE Sens J 11(12):3426–3432
39. Laenge K, Rapp BE, Rapp M (2008) Surface acoustic wave biosensors: a review. Anal Bioanal Chem 391:1509–1519
40. Lucklum R, Hauptmann P (2000) The quartz crystal microbalance: mass sensitivity, viscoelasticity and acoustic amplification. Sens Actuators B 70:30–36
41. Martin SJ, Frye GC, Senturia SD (1994) Dynamics and response of polymer-coated surface acoustic wave devices: effect of viscoelastic properties and film resonance. Anal Chem 66(14):2201–2219
42. Martin SJ, Ricco AJ, Niemczyk TM, Frye GC (1989) Characterization of SH acoustic plate mode liquid sensors. Sens Actuators 20:253–260
43. Martin SJ, Frye GC, Spates JJ, Butler MA (1996) Gas sensing with acoustic devices. In: IEEE ultrasonic symposium. San Antonio, Texas, pp 423–434
44. Morrison SR (1994) Chemical Sensors. In: Sze (ed) Semiconductor sensors. Wiley, New York
45. Pijanowska DG, Torbicz W (2005) Biosensors for bioanalytical applications. Bull Pol Acad Sci Tech 53(3):251–260
46. Rayleigh (Strutt JW) (1885) On waves propagated along the plane surface of an elastic solid. Proc Lond Math Soc 17:4–11
47. Ricco A, Martin S (1987) Acoustic wave viscosity sensor. Appl Phys Lett 50(21):1474
48. Ricco AJ, Martin SJ, Zipperian TE (1985) Surface acoustic wave gas sensor based on film conductivity changes. Sens Actuators 8:319–333
49. Sadek A, Baker Ch, Powell D, Włodarski W, Kaner R, Kalantar-Zadeh K (2007) Polyaniline nanofiber based surface acoustic wave gas sensors—effect of nanofiber diameter on H_2 response. IEEE Sens J 7(2):213–218
50. Sadek A, Włodarski W, Shin K, Kaner RB, Kalantar-Zadeh K (2008) A polyaniline /WO_3 nanofiber composite-based ZnO/64^0 YX $LiNbO_3$ SAW hydrogen gas sensor. Synthetic Met 158:29–32

51. Sauerbrey G (1959) Verwendung von Schwingquarzen zur Waegung dunner Schichten und zur Mikrowaegung. Z Phys 155:206
52. Sivaramakrishnan S, Rajamani R, Smith CS, McGee KA, Mann KR, Yamashita N (2008) Carbon nanotube coated surface acoustic wave sensor for carbon dioxide sensing. Sens Actuators B 132:296–304
53. Skłodowska-Curie M (1895) Radioactive substances. Dissertation, Philosophical Library Inc., New York, 1961
54. Śliwiński A (1993) Ultradźwięki i ich zastosowania (Ultrasounds and their applications). WNT, Warszawa
55. Soluch W (1998) Design of SAW delay lines for sensors. Sens Actuators A 67:60–64
56. Thomson M, Stone DC (1997) Surface-launched acoustic wave sensors. Wiley, New York
57. Urbańczyk M, Jakubik W (2000) Studies of the electrical and mass effects in surface acoustic wave gas sensor. Electron Technol 33(1):161–166
58. Urbańczyk M (2011) Sensory gazu z akustyczną falą powierzchniową (Gas sensors with surface acoustic wave). Silesian University of Technology, Gliwice
59. Vellekoop MJ (1998) Acoustic wave sensors and their technology. Ultrasonics 36:7–14
60. Wang W, Lee K, Woo I, Park I, Yang S (2007) Optimal design on SAW sensor for wireless pressure measurement based on reflective delay line. Sens Actuators A 139:2–6
61. Wang W, Shitang H, Shunzhou L, Minghua L, Yong P (2007) Enhanced sensitivity of SAW gas sensor coated molecularly imprinted polymer incorporating high frequency stability oscillator. Sens Actuators B 125:422–427
62. White R (1970) Surface elastic waves. Proc IEEE 58(8):1238–1277
63. White RM, Voltmer FW (1965) Direct piezoelectric coupling to surface elastic waves. Appl Phys Lett 7(12):314–316
64. White RM, Wicher PJ, Wenzel SW, Zellers ET (1987) Plate-mode ultrasonic oscillator sensors. IEEE Trans UFFC 34(2):162–171
65. Wohltjen H, Dessy R (1979) Surface acoustic wave probe for chemical analysis. Anal Chem 51:458–1470
66. Wrotniak J (2022) Akustyczne fale powierzchniowe w badaniach własności sensorowych cienkich warstw wybranych związków polimerowych z wykorzystaniem dodatkowych metod aktywacji (Surface acoustic waves in studies of sensor properties of selected polymer compounds thin films with the use of additional activation methods). Dissertation, Silesian University of Technology, Gliwice
67. Yatsuda H, Naro M, Kogai T, Lizawa H, Kurosawa S (2007) STW gas sensors using plasma-polymerized allylamine. Thin Solid Films 515:4105–4110

Chapter 3
Theoretical Models of SAW Sensor's Operation

Abstract This chapter contains the general models of the SAW sensor operation along with their mathematical description. The general model is based on three basic couplings occurring in gas sensors: gas molecules-sensor structure, then sensor structure-SAW propagation and measurement of altered SAW parameters by a suitably sensitive detection system. The mathematical description takes into account the influence of the most important parameters of the layers and sensor structures on the SAW propagation. This description adopts a typical division into the mass effect, resulting from the mechanical parameters of the sensor structure, and the acoustoelectric effect, which is a consequence of its electrical parameters, mainly electrical surface conductivity. For the first time, a detailed theoretical description of the acoustoelectric effect for bilayer sensor structures was presented.

Chapter 2 presents the general characteristics of SAW gas sensors, giving in the end their generally accepted design, utility characteristics and possible applications. The general principles of operation of this type of sensors, together with their mathematical models, will now be discussed.

From an utility point of view, a gas sensor, including SAW one, as a result of contact with the atmosphere in which the concentration of a particular gas exceeds the limit of its detection, makes it possible to determine the actual concentration. This is the final stage of operation of each sensor in the form of a measurement result. However, this result, depending on the type of sensor, is the result of several intermediate stages. As already mentioned, the most important and basic stage is the interaction between the molecules of a given gas and the sensor element used in a given type of gas sensor. From the point of view of sensor applications, the aim is that this element has sufficient sensitivity and selectivity with respect to the selected gas/vapour molecules. There must be changes in certain parameters of the sensor element due to the presence of gas. The values of these changes for small concentrations, often very small, are then precisely determined using appropriately sensitive measurement methods.

W. Jakubik, *Application of Surface Acoustic Waves in Gas Sensors*, Springer Series in Surface Sciences 69, https://doi.org/10.1007/978-3-031-89971-3_3

3.1 General Model

The essence of the gas sensor model is to represent the sorption of vapor or gas particles from the surrounding atmosphere by an appropriately made and matched sensor element. As a result of these physicochemical phenomena of adsorption and/or absorption, there are changes in the parameters of the sensor element, mainly its mass, its elastic constants, and electrical conductivity. The task of the transducer, on the other hand, is to convert these changed parameters into an electrical signal, depending on the concentration, more or less proportional to the amount of gas molecules absorbed. This simplest general model of operation of the gas sensor, consisting of only two parts, i.e. the sensor element and the transducer, is presented symbolically (Fig. 3.1) [61].

In the case of SAW sensors, the sensor elements are structures consisting of single thin layers and multilayer systems, including bilayer structures, made, depending on the purpose, of appropriate materials. The transducer, on the other hand, is partially the acoustic surface wave itself, whose basic propagation parameters, such as velocity and/or attenuation, are disturbed as a result of the changed parameters of the sensor element. The other parts of the transducer are the piezoelectric substrate, on which the surface wave spreads, as well as a suitably sensitive measurement method, allowing the determination of changes in velocity v, amplitude A, or wave phases ϕ (Fig. 3.2).

It follows from the presented description that in addition to the interactions between the sensor element and the gas molecules in its immediate environment, there is another interaction, between the sensor element and the propagating surface

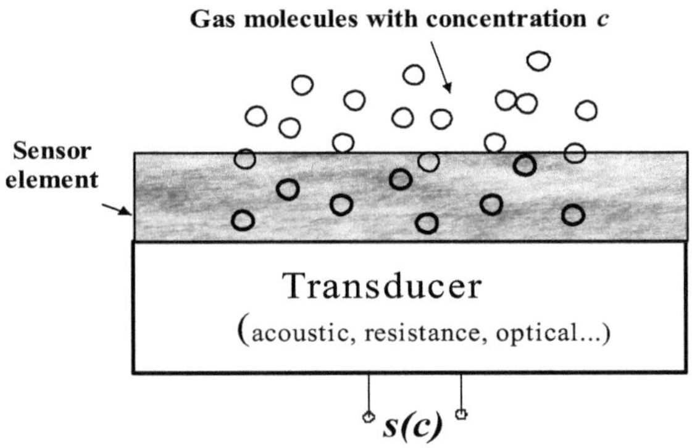

Fig. 3.1 The general model of the gas sensor: output signal s is a function of gas concentration c in the surrounding atmosphere

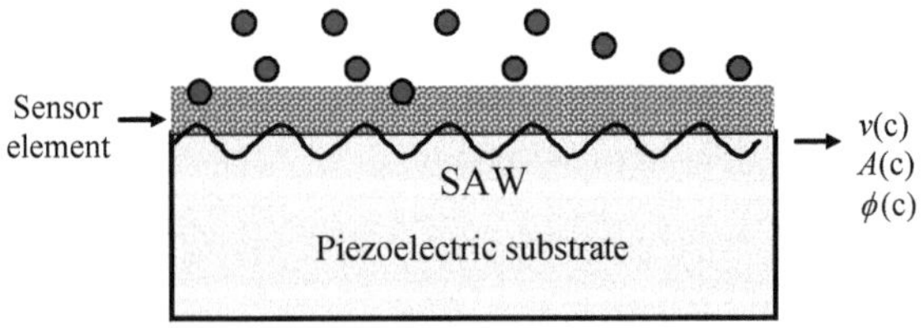

Fig. 3.2 SAW with sensor element and piezoelectric substrate as a transducer in gas sensor

wave. One can speak of a certain type of coupling, because the first interaction (gas molecules-sensor element) affects the second interaction (sensor element-SAW propagation). A generalization of the above model should therefore be presented in the category of couplings [28].

3.2 Basic Couplings in Gas Sensors

From the point of view of sensor applications, it should be remembered that the change in the basic parameters of the SAW, i.e. velocity, amplitude, or phase, is a secondary effect that is caused by the primary interaction between the sensor element and specific vapor or gas particles. This coupling determines the possible sensory applications of the layer or the entire structure. As a result of its existence, the parameters of the sensor element change, which determines the selectivity and partially the sensitivity of the entire sensor. Historically, the first coupling of this type was observed by Brattain and Bardeen [10], while working on the design of the transistor. The authors attributed certain instability to the behavior of the constructed device to gas adsorption on the German surface.

Another coupling that can be distinguished exists between the sensor element and the propagating surface wave. The capabilities of this coupling determine how much the altered parameters of the sensor element used can disturb the wave propagation. As already mentioned, the high sensitivity of the SAW to changes in the boundary conditions of its propagation allows a high sensitivity of the sensor. Graphically, these two basic couplings, found in most gas sensors, supplemented by the third coupling, i.e. the analysis of signals and the conversion of them into concentration, are shown in Fig. 3.3.

The most important is the first (**I**) coupling between the gas molecules and the sensor element. It decides on further couplings, i.e. if it is not present, no further steps can occur. At present, the most popular sensor elements can be thin, thick, ceramic layers, intensively studied carbon nanotubes (single-walled and multi-walled carbon nanotube (SWCNT), (MWCNT)) [71], as well as palladium nanowires [17], and layered structures [30, 36].

The effect of coupling (**I**) should be to modulate the parameters (mechanical, electrical, optical) of the sensor element under the influence of even trace amounts of selected molecules from the surrounding atmosphere. It should be noted that it is most difficult to determine small amounts of molecules at the level of single *ppm* or even *ppb*, which is related to the detection limit of available sensor systems.

However, just as important as the coupling (**I**) is the coupling (**II**), between the altered parameters of the sensor element and the detection system. The point is that this system must be sensitive to changes in the parameters of the sensor element, caused by the presence of molecules of selected gases. A distinction can be made between direct (e.g. resistance sensors) and indirect (sensors with SAW) detection schemes. In gas sensors with SAW, indirect detection occurs because changes caused by gas molecules (e.g. change in mass or electrical conductivity of the sensor element)

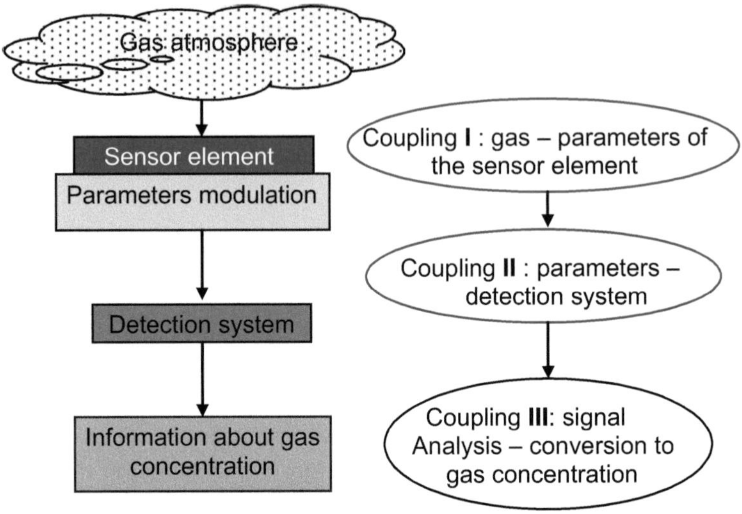

Fig. 3.3 The three basic couplings in gas sensors

are not measured directly. The measurement is made indirectly, by changing the wave velocity, which in turn is observed as frequency shifts in the oscillator system or by measuring a phase change. The important features of the measuring system will be discussed in Chap. 4.

Coupling (**III**) is the last step in the interaction processing sequence and concerns the analysis of the signals obtained. Various methods are used to correct the effects of long-term thermal drift, humidity of the gas atmosphere, and aging effects of sensor structures on the sensor signal. The ultimate goal is to obtain useful information about the concentration of the selected gas. This is a very important coupling, but without proper couplings (I) and (II) it is impossible to carry out.

The next part of the chapter is devoted to the description of the coupling (I) between the gas molecules and the parameters of the sensor element, with consideration of the kinetics of the phenomena occurring, and the coupling (II) between the parameters of the sensing element and the detection system, which is the SAW Rayleigh type. The results of tests of selected sensor structures, as well as attempts to analyze the signals obtained in the form of changes in frequency in function of time and concentration of gases, i.e. coupling (III), are included in Chap. 5.

3.3 Kinetic Phenomena for the Interaction Gas-Sensing Element

In the case of gas sensors with SAW, the sensor elements are primarily thin-film structures. Essentially, the interaction process between molecules of any gas or vapor and the thin-film sensor structure is a kinetic phenomenon, with at least three steps (Fig. 3.4).

The first stage is the transport of the gas molecule to the surface of the sensor material, where it can undergo physisorption or chemisorption. The second stage is the diffusion into the interior of the sensor material; this can interfere the SAW propagation. The last stage is the return of the molecule to the surface and its desorption to the surrounding atmosphere. Of course, only surface adsorption can occur without diffusion to the interior of the layer. Each of these cases can be distinguished as (1) the process of the molecule being transported to the surface, (2) the finite time remaining on the surface or in the volume of the layer, and (3) the phenomenon of desorption into the surrounding atmosphere.

Important variables responsible for kinetic phenomena are the diffusion properties of the given gas molecules, their interaction energies with the sensor structures, and the solubility parameters of the gas in a given layer. In addition, each of these variables will depend on the temperature at which the interaction occurs. The gas flow velocity in dynamic studies can also affect the kinetics of the interaction phenomena between the gas molecules and the sensor elements used [80].

A substance in the gas phase has a higher entropy than one adsorbed on the surface or dissolved in the volume of the sensor element. Therefore, there must be an affinity (for that substance in the gas phase, relative to the surface or volume of the sensor element), which is manifested by a decrease in the "binding energy" of the substance that has been absorbed. One possibility of adsorption of molecules on the surface of the sensor element is to "trap" them. At steady state, the concentration of the adsorbed substance in the sensor element c_{se} relative to its concentration in the surrounding gas phase c is determined by the *partition coefficient* K [5]:

$$K = \frac{c_{se}}{c} = e^{-(\Delta H_g - T\Delta S_g)/RT} \tag{3.1}$$

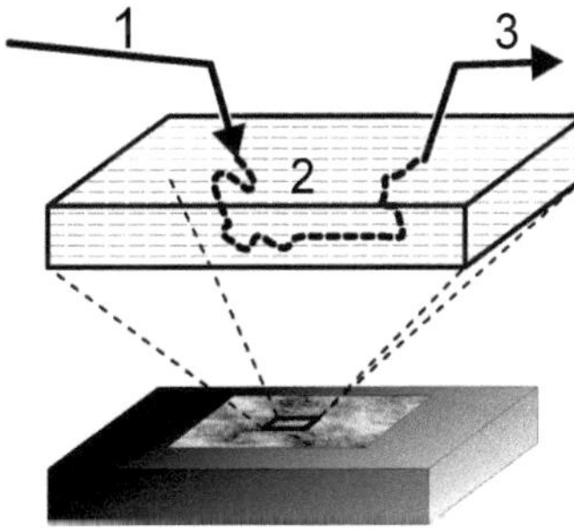

Fig. 3.4 Representation of the molecule transport to the sensor element surface (1), sorption and diffusion to inner part of the sensor (2), and recovery desorption to the surrounding atmosphere (3) [80]

where ΔH_g and ΔS_g denote changes in enthalpy (binding energy) and entropy, respectively, for the gas sorption phenomenon that occurs, R is the gas constant, and T is the reaction temperature in absolute scale. From Eq. (3.1), it follows that for a substance that undergoes sorption, the binding energy (ΔH_g) must be sufficient to overcome the entropy loss (ΔS_g) for the transition from the gas phase to the condensed in the sensor structure. In addition, if the interactions become stronger (more negative ΔH_g), the number of adsorbed molecules at a given concentration in the gas phase will increase. This will increase the sensitivity and lower the detection threshold.

From the point of view of the detection of different substances, it seems that stronger interactions between molecules and sensor elements are always beneficial. However, there is an optimal state of sensitivity of the sensor and the processes of gas sorption. The average residence time, when the adsorbed molecule remains on the surface, is [1]

$$\tau_r = \tau_{r0} e^{-\Delta H_g / RT} \tag{3.2}$$

where τ_{r0} is on the order of 10^{-12} to 10^{-13} s. Therefore, when the concentration of the substance in the gas phase changes, the time to reach the new equilibrium state increases, as well as the detection of the change in this concentration requires longer times for higher binding energies.

The phenomenon of the sorption of a particle of any substance requires it to reach the sensor element (Fig. 3.4). In the case of sensors in the form of thin layers, the molecule is adsorbed on the surface. The strength of the binding then depends on the nature of the physical and chemical interactions between the molecules of the substance and the surface of the sensor element. This process can take place using weak physical interactions (*physisorption*) of the van der Waals type, for which the change in binding energy of ΔH_g is less than ~40 kJ/mol. Such interactions are easily reversible and equilibrium is achieved fairly quickly ($\tau_r \sim 10^{-12}$ s). In the case of chemical interactions (*chemisorption*), we are dealing with covalence or ionic bonds with a large change in binding energy $\Delta H_g \sim$ 80–400 kJ/mol.

In the case of physisorption, temperature has a very large influence on the responses of the sensor structures, by changing the rate of both adsorption and desorption. In addition, the sensor response is also influenced by the possible surface area of the impact and the surface density of the so-called *active states*—places where the adsorbed molecule can bind to the surface. Low interaction temperatures and increasing surface area (porous layers) allow for greater sensitivity and lower detection limits, at the expense of longer response and regeneration times. The issue of selectivity for this type of interactions is related to the stronger or weaker affinity of the substance to be detected relative to the sensor material. Increasing the surface temperature of the sensor layer can result in low affinity substances not condensing on the sensor surface.

The physical and chemical nature of the sensor structures is reflected in the adsorption isotherms for the analyzed substances. The simplest model of gas adsorption in the form of a monolayer on the surface of the sensor element is the Langmuir isotherm. In this kinetic model, it is assumed that the molecules hitting the surface of the sensing structure undergo adsorption at precisely defined locations—so-called *active centers (states)*. Only one molecule can occupy one state at a time and there are no internal interactions between molecules adsorbed in adjacent active states. Equilibrium particle coverage for surface active states θ is achieved for a certain partial pressure p' in the surrounding atmosphere [1]:

$$\theta = \frac{Bp'}{1 + Bp'} \tag{3.3}$$

where the coefficient B is related to the relative rate of adsorption and desorption and the number of particles hitting the surface that are trapped.

This equation means that at low concentrations in the environment of the sensor structure, the surface coverage and the sensor response are linear functions of concentration. In contrast, at high concentrations, the response becomes saturated, which corresponds to the coverage of the sensor structure by a single monolayer. Therefore, the range of concentration at which the sensor reacts depends on the equilibrium constant B, which in Eq. (3.3) can be roughly estimated on the basis of kinetic gas theory [80]:

$$B = \frac{h'^3}{k_B T (2\pi m k_B T)} \frac{f_a(T)}{f_g(T)} \exp\left(\frac{E_{ag}}{k_B T}\right) \tag{3.4}$$

where $f_a(T)$ and $f_g(T)$ are the internal participation functions for the molecule in the adsorbed and gas states, respectively; E_{ag} is the energy required to move the molecule from the lowest adsorbed state to the lowest state in the gas phase; h' is Planck's constant; k_B is Boltzmann's constant; and m is mass of the gas molecule.

On the other hand, the additional mass per unit area (surface density) resulting from adsorption is

$$\Delta\rho_{pow} = N_s \theta m_1 \tag{3.5}$$

where N_s is the density of surface active centers (typically it is ~10^{15} cm^{-2}) and m_1 is the mass of a single adsorbed molecule.

Another problem in the case of physisorption is the possibility of producing multiple layers by adsorbed gas molecules. Such phenomena have been classified in the category of five types of isothermal adsorption [11], which is important for the response and regeneration times of gas sensors. The most important, however, are changes in the parameters of the sensor elements used.

3.4 Models of Coupling Gas-Sensor Element

The interactions between gas molecules or vapors of organic compounds and sensor elements are complex physicochemical processes. Depending on the type of element used (e.g. thin, thick layers, layered systems) and the type of materials used (metallic layers, semiconducting metal oxides, organic semiconductors), these processes can be more or less complex. The primary reason for these processes is that the forces acting in the volume of the material, responsible for the coherence of its crystal lattice, are only partially saturated at the surface. As a result, the ability to adsorb vapors and gases on the surfaces of many materials is observed [8].

General models of interactions between gas molecules and the most commonly used sensor elements such as metal oxides, metalophthalocyanines, palladium thin layers, and metalophthalocyanine layer structures will be presented. The magnitude of these interactions is measured by changes in the parameters of the sensor elements. Proper identification of the sensor mechanism of this basic coupling is extremely important, as it can improve the performance of the gas sensor.

3.4.1 *Metal Oxides*

It has long been known that the concentration of electrical charge carriers near the surface of semiconductor metal oxides changes under the influence of the presence of certain gas molecules. The first practical use of this phenomenon was made by Seiyama et al. [70], for thin layers of ZnO at a temperature of about 300 °C, by measuring the change in resistance in the function of gas concentration. Later work focused on tin dioxide SnO_2, which was also characterized by greater stability [78]. Since then, there has been a very large development of gas sensors using tin dioxide, the most famous of which are the *Taguchi Gas Sensor* (TGS)-type sensors, which are mass-produced.

Resistance sensors based on n-type semiconductors work in the air atmosphere as a reference gas. The oxygen always present in the atmosphere is absorbed on the surface of metal oxide in the form of O_2^-, O^-, and O^{2-}. Which form prevails depends on the temperature. Most often it is assumed that O^- dominates. In order this form to be formed as a result of absorption, a reaction is proposed:

$$O_2 + 2e^- \rightarrow 2O^- \tag{3.6}$$

As a result of this reaction, some of the electrons are bound on the surface with oxygen, which results in a reduction in the electrical conductivity of the material.

In the presence of reducing gas (e.g. hydrogen) there is a reaction:

$$H_2 + O^- \rightarrow H_2O + e^- \tag{3.7}$$

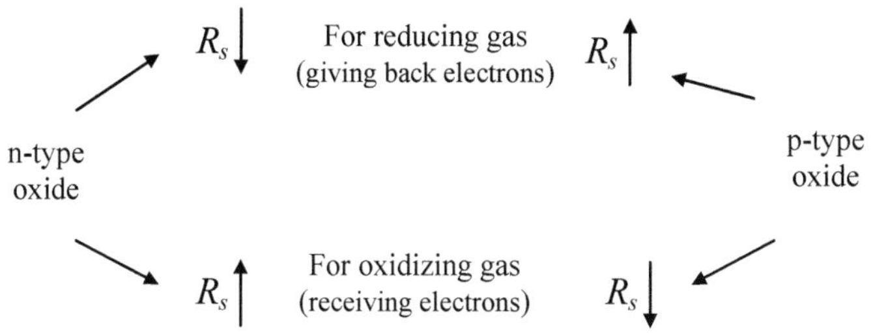

Fig. 3.5 Increase and decrease of the *n*- and *p*-type semiconductor metal oxide sample resistance as a result of interaction with reducing and oxidizing gases

This results in electrons being released. In the entire balance, this is manifested by an increase in electrical conductivity [47].

The presented interaction scheme between gas molecules and semiconducting metal oxides, such as SnO_2, was proposed by Windishmann and Mark [87] and subsequently repeated in many papers (e.g. [64]). This model is simplified because it does not take into account the presence of water vapor as well as the influence of the morphological structure of the sensor materials.

In the case of a semiconductor with n-type conductivity, interaction with reducing gases (e.g. CO, CH_4, H_2) results in an increase in electron concentration, i.e. a decrease in the resistance of the sample R_s. On the other hand, interaction with oxidizing gases (e.g. NO_2, Cl_2, O_3) results in the opposite effect, i.e. an increase in the resistance of a sample [86]. For materials with p-type conductivity which are rare (e.g. NiO, CuO) [13], the behavior is reversed (Fig. 3.5).

At high temperatures in metal oxides that are then stable (e.g. TiO_2), the mechanism changes from surface to volumetric. As a result of interaction with the reducing gas, point defects are formed, including oxygen vacancies, whose ionization leads to an increase in the concentration of carriers in the volume, assuming that the oxygen pressure is constant for the air. In general, it can be stated that in this model, the sample resistance is inversely proportional to the concentration of oxygen vacancy, which in turn is a function of the concentration of the molecules of the given gas [64].

Based on the presented interaction models, it can be concluded that semiconductor metal oxides as sensor elements are characterized primarily by high sensitivity to change in oxygen concentration in air. Resistive gas sensors require a high operating temperature (SnO_2 ~ 350 °C). However, sensor elements made of metal oxides exhibit poor selectivity, which is compensated by their low price.

Currently, the sensor elements used in gas sensors are mainly made of SnO_2 in thick-film form (e.g. [55]) and thin-film form (e.g. [75]). The SnO_2-Rh (rod) layer structures with thicknesses of ~40 μm are also used [56]. They can be used for methane detection, the operating temperatures are then in the range 400–600 °C. One of the first sensor materials, i.e. ZnO in the scope of methane detection is returning, the working temperatures are then lower ~100–300 °C (e.g. [63]).

ZnO layers, as piezoelectrics, are also used in SAW sensors. Therefore, it is possible to make this type of sensor on a non-piezoelectric substrate, such as glass. The wave then propagates in the ZnO layer, which serves a dual role as a piezoelectric waveguide and as a sensor element [18].

The TiO_2-SnO_2-layered systems (e.g. [90]) and WO_3-layered structures with Pd, Au, Bi, Sb (e.g. [67]) are also used in resistance sensors. It is also interesting to use the dynamically changing temperature of a sensor element made of SnO_2 [62]. Layered sensor elements using metal oxides are also used in SAW sensors, which are characterized by increased sensitivity, as a result of the higher wave energy density at the propagation surface. Examples are WO_3-Pt or Au structures for hydrogen detection [22], as well as for the first time used bilayer structures WO_3-Pd, characterized by very high sensitivity with respect to hydrogen than single sensor elements [25, 27].

The work on semiconductor metal oxides in gas sensors is extremely extensive, which is why only selected groups from European centers will be listed. The group in Tuebingen (e.g. [6]), while in terms of nanostructures, groups in Brescia (e.g. [12]) and Barcelona (e.g. [3]).

3.4.2 Phthalocyanines

In addition to metal oxides, organic semiconductors, such as metalophthalocyanines, which belong to polymeric sensor materials, are used as material for sensor elements. When using thin layers of macromolecular compounds (e.g. metalophthalocyanine), a phenomenon similar to that of metal oxides is observed, i.e. their electrical conductivity strongly depends on the presence of certain gas molecules, even at very small concentrations at the level of single *ppm*. The change in electrical conductivity can be many times greater than that of metal oxides. The interaction of gas molecules with both the positive ion of the metal in phthalocyanine and the electron cloud at the periphery of its macromolecules stands out.

In this first case, the positive metal ion at the center of the molecule interacts with the electropositive gases to form a *coordination complex* (*CC*), while in the second case, a large electron cloud of the outer benzene rings interacts with electronegative gases to form a *charge transfer complex* (*CTC*) (Fig. 3.6) [9]. It is now believed that gases such as CO and NH_3 form coordination compounds when interacting with phthalocyanines, while NO_2 is an example of a charge transfer relationship. An electron affinity of 3.6 eV for NO_2 and less than 1 eV for other gases is an important parameter for these interactions [65].

The type of central metal ion in the metallophthalocyanine complex (MPc) molecule has a great influence on its physicochemical properties. In particular, the oxidative and reduction properties are changed, which is important for the adsorption phenomenon of certain gas molecules on the surface of thin layers of such materials. By adding various additional functional groups to the macrocyclic ring of the MPc molecule, its sensory properties can be further altered (e.g. [73]).

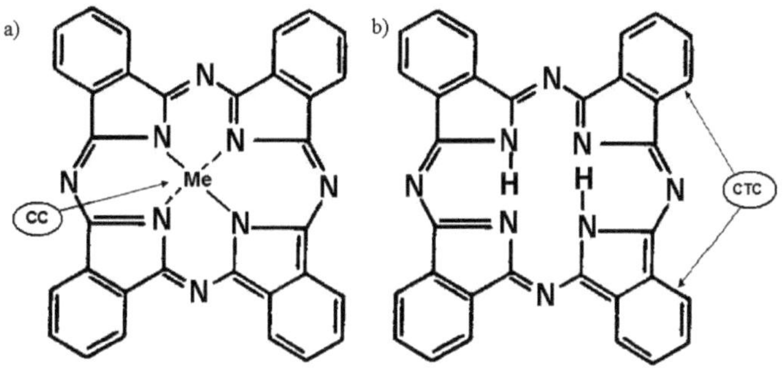

Fig. 3.6 CC-type interaction for metal complex phthalocyanine (**a**) and CTC-type interaction for its metal-free complex (**b**)

The advantage of thin layers of organic semiconductors of the phthalocyanine type is the expanded surface area of this structure, which leads to a higher density of surface active centers (Fig. 3.7). This greatly increases the absorption capacity of this type of sensor element through a higher number of active centers. The much lower operating temperature for this type of material (~70 °C) compared to metal oxides (~300 °C) is also very important. This allows to significantly reduce the energy consumption during practical use. However, a disadvantage is the poor desorption, e.g. for NO_2. The interaction of 83-nm-thick lead phthalocyanine (PbPc) with NO_2 with a very low concentration of ~0.64 ppm in air at ~70 °C resulted in a change in the differential frequency of the SAW system by up to 3 kHz [41]. In the same work, a method of accelerated desorption was proposed by repeated cooling and heating of the structure ("thermal pumping") (see Fig. 5.3).

Sensor properties of thin layers of various phthalocyanine types have been used in gas sensors for many years in numerous foreign centers (e.g. [66, 89]) as well as in Poland (e.g. [20, 76, 82]). On the other hand, the use of these materials in SAW sensors, e.g. in the form of layered sensor elements of the MPc-palladium, favors the achievement of high sensitivity of this type of sensor elements, at the same time as a satisfactory response rate [23]. The strongly expanded surface of phthalocyanine forces the "island" rise of palladium applied to its surface. In SAW gas sensors with layered sensor structures, the ability to achieve high sensitivity is also the result of some additional possibilities of interaction of these structures with the surface wave. This issue will be discussed in the next section (see Sect. 3.5) on the coupling of the sensor element-SAW, while the properties of thin palladium layers are currently presented, which are of great importance in the field of hydrogen detection.

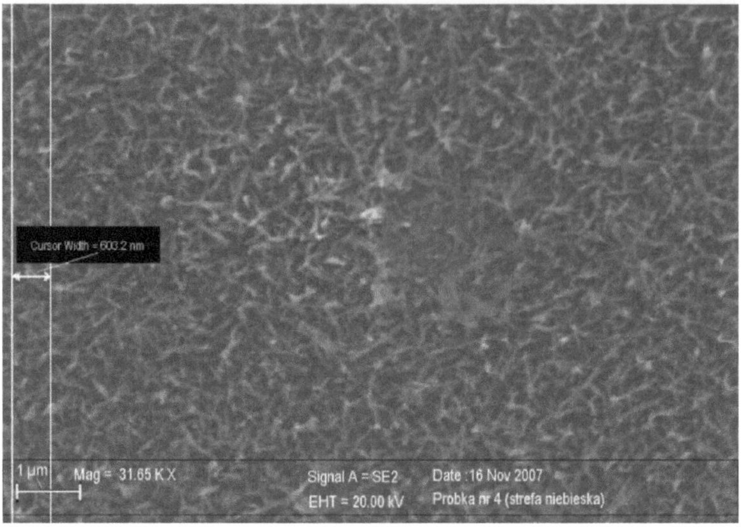

Fig. 3.7 SEM image of metal-free phthalocyanine ~110 nm used in investigations before covering by palladium

3.4.3 Thin Palladium Layers

Palladium has long been known as an excellent material for storing hydrogen [54]. Information provided by Lewis indicates that as early as 1866, palladium was used by Thomas Graham to absorb hydrogen during electrolysis. Since then, the use of palladium has been very common both in industry and more recently in sensing technology in hydrogen sensors.

In gas sensors (e.g. MOS and SAW), palladium thin layers perform three basic functions: they facilitate the dissociation of hydrogen molecules through their catalytic action, they transport hydrogen atoms to the interphase surface (e.g. phthalocyanine-palladium), and they adsorb hydrogen atoms to this interphase surface in the form of electric dipoles [47].

In the SAW sensor, a thin layer of palladium, due to its metallic nature, has an additional function, increasing the electrical conductivity of the layered structure. In many cases, this allows for strong acoustoelectric interactions that far exceed typical mass interactions.

The catalytic function means that hydrogen molecules and other hydrogen-containing compounds dissociate, forming single-atom hydrogen, which then diffuses into the interior of the palladium. Palladium oxide also has catalytic functions, which is important if it is produced on the surface of Pd. Other catalytic reactions are also possible, for example, for oxygen with hydrogen.

The second function is the transport of hydrogen single atom through the metal layer. Hydrogen is unique in this case because of its extremely small ion radius compared to other atoms. No other atom is able to diffuse through the metal layer so quickly. The diffusion ratio of hydrogen to the volume of palladium is estimated at D_0 ~2.83×10^{-7} m^2/s with an activation energy of ~5.39 kcal/mol [49]. In the case of porous materials, diffusion is much easier, even without the participation of catalytic dissociation.

The third function of the palladium layers is the adsorption of hydrogen atoms that diffuse into the layer on the inner surface, as electric dipoles. The dipole layer formed at this location will cause a change in the electric field on the surface of the semiconductor (both organic and inorganic) on which the palladium layer was formed. This additional electric field can change the distribution of charges in the semiconductor, and thus its electrical conductivity, which as a result can lead to a disturbance in the propagation rate of SAW through an acoustoelectric phenomenon.

Under the influence of hydrogen, both mechanical and electrical properties of the thin layer of palladium change. The change in mechanical parameters is, among other things, a consequence of the incorporation of hydrogen into the structure of palladium. It is assumed that 350 to 850 volumes of hydrogen can be incorporated into a single volume of palladium. The ability to form a diamagnetic compound PdH or PdH^+ can be considered to be the reason for such a high absorption capacity of palladium against hydrogen, which causes a decrease in the magnetic sensitivity of Pd as it becomes saturated with hydrogen [72].

The hydrogenated palladium layer (PdH_x) occurs in two solid forms: α and β. The α phase is a reversible structure and is formed as a result of interaction with low hydrogen concentrations. As the hydrogen content increases, the α form changes into β form. This phenomenon has been known for a very long time and has been investigated by Owen and Jones (1937). The H_2 concentration at which the α-phase change to β occurs depending on the thickness of the palladium layer and temperature. Based on literature data [77], it can be stated that the moment of the α- to β-phase change occurs for a palladium layer with a thickness of 10–100 nm, at a concentration of about 2% hydrogen at a temperature of about 30 °C. The β-PdH_x phase has a regular structure like Pd. Palladium alloys with chromium, silver, or nickel are used to prevent the α phase from turning into β [79].

The maximum hydrogen concentration, which does not yet cause a phase transition, shifts towards higher concentrations. Phase stabilization can also be ensured by increasing the temperature of the interaction [36]. At high hydrogen concentrations, the Pd layer is mechanically unstable.

Among the most recent achievements that exploit the absorption properties of palladium relative to hydrogen are the matrices of palladium nanowires. These structures exploit the expansion or contraction of the Pd nanowire crystal lattice under the influence of hydrogen. This makes it possible to obtain a very fast response because such structures act almost like switches with a reaction time of ~75 ms [17, 84].

3.4.4 Layered Structures

As mentioned in Chap. 1, the excellent absorption properties of single thin palladium layers cannot be exploited in SAW sensors. The resulting mass effects for very light hydrogen atoms are negligible, even for relatively large palladium layer thicknesses [15]. Therefore, the concept of semiconductor-palladium layer structures, e.g. metalophthalocyanine-palladium (MPc + Pd) or metal oxide-palladium (MO + Pd), was proposed, for which the model of interaction with hydrogen is presented in Fig. 3.8. This model was adopted for SAW hydrogen sensors based on the work of Lundstroem et al. [57] on MOS sensors.

The most important thing is to create a dipole layer on the surface connecting the palladium to the semiconductor. The additional electric field changes the electrical conductivity of the semiconductor layer, which in turn can cause a change in the propagation velocity of the SAW, and ultimately a change in the frequency of the oscillation (for the oscillator system).

Theoretically, the difference in potentials resulting from the formation of such a dipole layer $\Delta\varphi$ will be [74]

$$\Delta\varphi = \pm\frac{\mu_H N_s \theta}{\varepsilon_0} \tag{3.8}$$

where μ_H is the dipole moment of the adsorbed hydrogen atom, N_s is the density of the adsorption states, θ is the degree of surface coverage, and ε_0 is the electrical permeability of the vacuum.

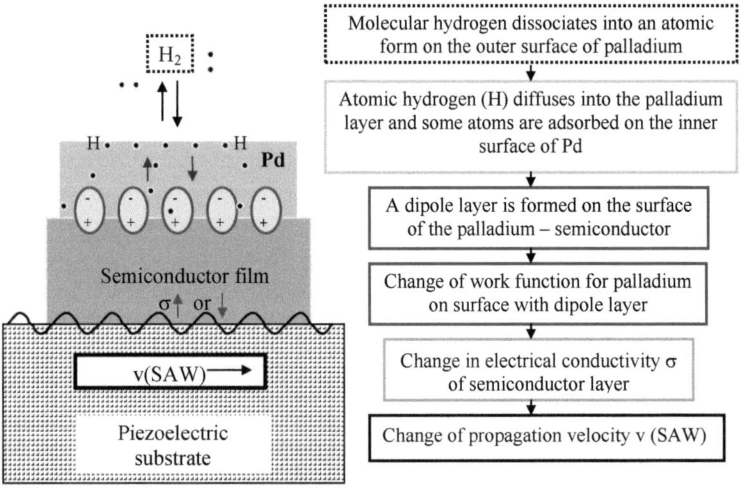

Fig. 3.8 Model of the bilayer structure (MPc-Pd and MO-Pd type) interaction with hydrogen in SAW sensor [33]

When hydrogen is adsorbed on catalytic surfaces such as Pd, Ni, or Pt, two types of adsorption can be observed [58]. The *r-adsorption* type increases while the *s-adsorption* type decreases work function, hence the sign ± in Eq. (3.8).

Since the electrical dipole moment of adsorbed hydrogen atom is on the order of 2×10^{-31} cm, and the average surface density of palladium atoms $N_s =$ ~2×10^{19} m^{-2}, so the maximum potential difference for the semiconductor layer will be

$$\Delta\varphi_{\max} = \frac{\mu_H N_s}{\varepsilon_0} \cong 0.45\ V \tag{3.9}$$

The resulting electric field strength (~10^6 V/m for layers with a thickness of ~100 nm) is large enough to disturb the concentration and mobility of the charge carriers in the semiconductor layer. As a result, this causes a change in its electrical surface conductivity, which can cause a disturbance in the propagation velocity of the SAW as a result of the acoustoelectric interaction. The maximum magnitude of this disturbance is a change of ~$K^2/2$, or in the case of a $LiNbO_3$ Y–Z substrate a decrease of 2.4% in SAW velocity.

3.4.5 The Change of Sensing Element Parameters

From the point of view of the SAW propagation on the surface of a piezoelectric, changes in the parameters of the sensor structures performed on its surface are important. In the case of gas sensors, these changes must be the result of the interaction of the structure with vapor or gas molecules in its environment. Depending on the material used for the sensor structure, the mechanical, electrical, and viscoelastic parameters can change. While the first two groups of parameters are characteristic for thin layers of metallic and semiconducting materials (e.g. phthalocyanines, metal oxides), changes in viscoelastic parameters are characteristic for some polymer materials. The changed parameters of the sensor structures then affect the SAW propagation (see Sect. 3.5).

The thin layers of palladium change their mechanical parameters under the influence of hydrogen absorption. In particular, there is a decrease in their density (from 12.02 to 11.68 g/cm^3 for H_2 concentrations below 3%) and a change in their elastic modulus (usually also a decrease from 1.21×10^{11} to 1.04×10^{11} Pa for H_2 concentrations under 3%) [91].

Table 3.1 shows the relative changes in density $\Delta\rho/\rho_0$, elastic moduli $\Delta C_{11}/C_{11}$ and $\Delta C_{44}/C_{44}$ obtained at room temperature for thin layers of palladium made on SiO_2, for three concentrations of hydrogen mixed with nitrogen ($H_2 + N_2$) and the resulting relative changes in velocity $\Delta v/v_0$ of SAW.

Small changes in the density (0.11–0.65%) and elastic modulus (0.98–2.05%) of the palladium layers, resulting from hydrogen absorption, result in a reduction in the SAW velocity in the range of 5.6–18.9% for the structure subjected to long-term vacuum annealing (*). Significant changes in the elastic modulus, both reduction

Table 3.1 Relative changes in mechanical parameters of palladium thin layers in $H_2 + N_2$ mixture obtained at room temperature and relative changes in SAW velocity on SiO_2 substrate

Pd layer	H_2 concentration in %	$\Delta\rho/\rho_0$ in %	$\Delta C_{11}/C_{11}$ in %	$\Delta C_{44}/C_{44}$ in %	$\Delta v/v_0$ $\times 10^{-6}$
*	0.1	−0.33	−0.98	−1.0	−5.6
*	0.5	−0.65	−1.72	−2.05	−11.6
*	1.0	−0.64	−1.90	−2.05	−18.9
**	1.0	−0.11	−29.0	+28.0	+104

* Annealing in vacuum 10^{-7} mbar within 10 h, ** without annealing [2]

(−29% for C_{11}) and increase (+28% for C_{44}) at 1% hydrogen concentration, were found for the sample not subjected to the annealing process (**). Such large changes in the Pd layer elastic modulus result in a significant increase in the SAW propagation velocity ($+104 \times 10^{-6}$). Large changes in parameters (~30%) of palladium layers are the result of lack of structural stability, which can be achieved by long-term heating in vacuum.

Recent studies of this type include [16] and [91], who also found that for hydrogen concentrations below 3%, the density of the palladium layers decreased (by 2%), while Young's moduli decreased from 128 to 110 GPa (about 14%).

In addition to the changes in the mechanical parameters of the sensor structures as a result of interaction with gas molecules, there are also changes in their electrical conductivity. The author of the paper examined the electrical conductivity for selected layered sensor structures. Depending on the type of the tested structure, both a decrease (for CuPc + Pd) and an increase (for H_2Pc + Pd), for their electrical conductivity under the influence of hydrogen, of CO or water vapor in the air were observed. The temperature of the interaction is also an important parameter here. Table 3.2 presents a summary of the relative changes of this parameter for selected bilayer structures, based on the own research.

Table 3.2 shows that in addition to the bilayer structures with copper phthalocyanine, in the case of larger thicknesses (100 and 200 nm), the electrical surface conductivity increases as a result of interaction with hydrogen. The largest changes (+425%) were observed for the structure with hydrogen phthalocyanine at a thickness of 160 nm and a palladium layer at a low impact temperature of 30–31 °C.

A very small increase in electrical conductivity (+0.12%) was recorded for the NiO_x-Pd structure when it interacts with carbon monoxide in the air. On the other hand, an extremely large increase in electric conductivity (+800%) occurred for the Nafion-PANI (polyaniline) bilayer structure when the air humidity changed from dry to moist.

Relatively large relative changes in the surface electrical conductivity of the studied layer structures can cause acoustoelectric interactions that are much larger than typical mass interactions. It is important here to match the electrical conductivity of the entire layer structure to the active area of strong acoustoelectric interactions. For the defined structure, this adjustment is only possible to a small extent

Table 3.2 Ranges of relative changes in surface electrical conductivity $\Delta\sigma_s/\sigma_{s0}$ selected bilayer structures under the influence of hydrogen, CO, and humidity in the air

Layered structure, thickness in nm	Temperature, °C	H_2 concentrations' range %	Changes' range $\Delta\sigma_s/\sigma_{s0}$ %	Remarks from the own studies
200 CuPc + 18 Pd	34	0.01–0.08	0 to −10	No decrease σ_s for H_2 0.01%[a]
100 CuPc + 18 Pd	30–32	0.5–4	−58	Reductions σ_s almost uniform to concentrations[b]
100 CuPc + 18 Pd	38	0.5–4	−64	Reductions σ_s for H_2 concentrations above 2.5%[a]
80 CuPc + 18 Pd	33	0.5–4	0 to +18	Range of concentrations 1.5–3%[a]
160 H_2Pc + 20 Pd	30–31	0.5–4	0 to +**425**	Increase σ_s almost proportional to concentration H_2[c]
120 H_2Pc + 20 Pd	50	0.5–4	+53	Increase σ_s evenly for each H_2 concentration[c]
60 NiO_x + 18 Pd	35	1–2.5	~0	Changes σ_s at detection threshold[d]
50 WO_3 + 18 Pd	50	1–4	~+22	Incomplete saturation state in time 300 s[e]
60 NiO_x + 18 Pd	60	CO in ppm 50–100	+0.12	Both in dry and humid atmospheres[d]
300 Nafion + 72 PANI (polyaniline)	31	Change in humidity 5–50%	**+800**	Very large increase σ_s due to change in air humidity[f]

[a] [39], [b] [24], [c] [38] [d] [46, 40], [e] [25], [f] [26]

by changing the temperature of the interaction. On the other hand, to a large extent, it is possible to talk about structural fit by making bilayer structures of the semiconductor–metal type, in which there are changes in the resultant electrical conductivity even by several orders of size.

3.4.6 Response Profiles of SAW Gas Sensors

The SAW sensor system response is the result of several phenomena: adsorption, diffusion, and desorption. The rates of these phenomena (and the corresponding dynamic equilibrium state) are dependent both on temperature, sensor structure properties (thickness, porosity, etc.), and the concentration of the molecules to be detected. In other words, a change in the concentration of molecules in the c-gas phase gives a corresponding change in the concentrations of these molecules in the condensed phase c_{se} of the sensor element, ultimately producing the sensor response. In general, it can be expected that c_{se} will be a function of c, and these variables will not be equal. In some cases, when the sorption phenomena are irreversible, $c_{se} > c$.

For SAW sensor systems that are subject to the Langmuir isotherm, the expected dynamic response curve for a given concentration of molecules (assuming pure mass interactions) would look like Fig. 3.9a. The profile of this type, which is an example of an ideal response, clearly shows (1) fast initial adsorption until the formation of a monolayer on the surface and (2) much slower diffusion as a result of absorption phenomena into the volume of the sensor structure, e.g. polymer. When the factor causing the interaction disappears, the structure regenerates to its initial values, i.e. desorption. In the case of polymer coatings, the increase in temperature leads to an acceleration of the interactions (both faster response and regeneration), but at the expense of a lower concentration of the substance in the condensed layer phase, i.e. lower sensitivity and detection limit. However, this applies only to mass interactions. The ideal response of the sensor structure is symmetrical in the sense that the response curve is almost identical to the regeneration curve. In practice, however, it is difficult to fully regenerate the layer, and it takes much longer than adsorption.

In the case of interactions based on chemisorption mechanisms, the sensor response profile is often irreversible (Fig. 3.9b). An example of this type of response is shown in Fig. 3.10 for a thin layer of lead phthalocyanine with a thickness of ~310 nm and an NO_2 concentration of ~75 ppm in air and an interaction temperature of about 20 °C.

From the point of view of sensor applications, this is very unfavorable. One can try to raise the temperature, which often only slightly improves desorption. The "thermal pumping" proposed for the 83 nm PbPc layer turned out to be an effective solution (see Fig. 5.3). By repeatedly raising and lowering the temperature in the range of 70–80 °C, it was possible to achieve its reversible response in a time of ~1500 s at an air flow of 27.5 cm^3/s [41].

For some polymer coatings, as well as phthalocyanine-palladium layered structures, an interaction mechanism based on changes in viscoelastic parameters may occur. The SAW sensor response profile will then have a shape similar to that shown in Fig. 3.9c. Two phases of interaction can be distinguished here: rapid adsorption, i.e. mass interactions, and then a relaxing change in the elastic parameters resulting in an increase in frequency (Fig. 3.11).

The resulting relaxation changes are small, but clearly visible, especially since there are no changes in the resistance of the sample, which indicates the absence of

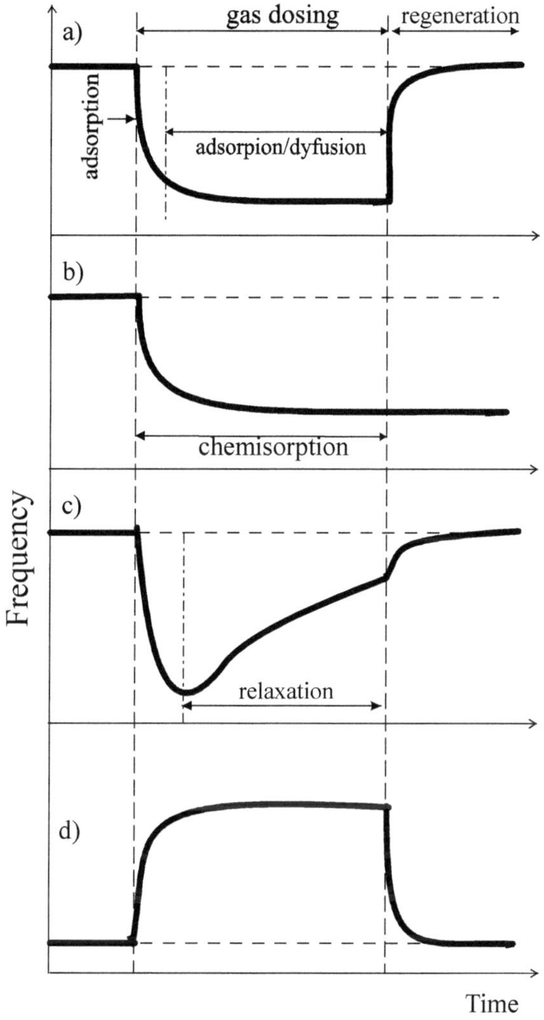

Fig. 3.9 Response profiles of the SAW sensor: **a** an ideal response, **b** irreversible chemisorption, **c** complex mass and viscoelastic interactions, **d** significant change in elastic properties of the sensor structure and acoustoelectric interactions [80]

acoustoelectric interactions. In addition, the instability of the differential frequency, caused by a small increase in the temperature of the sensor structure, is visible.

Response profiles (Fig. 3.9d) are generated for very large changes in the elastic parameters of sensor elements and for acoustoelectric interactions in layered sensor structures. Examples of such interactions are presented in a number of graphs in Chap. 5.

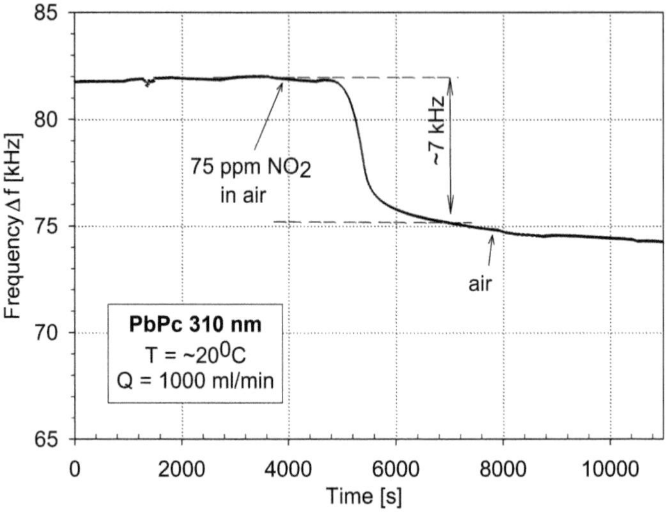

Fig. 3.10 Example of irreversible response (frequency decrease of ~7 kHz) for thin lead phthalocyanine (PbPc) film with thickness ~310 nm and after dosing of 75 ppm NO_2 in air at room temperature ~20 °C [28]

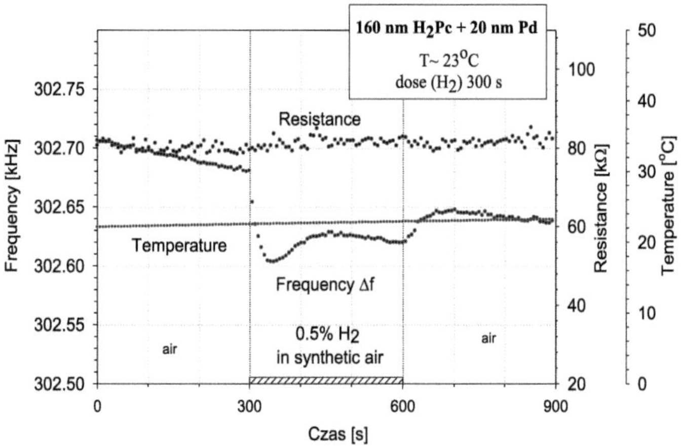

Fig. 3.11 Example of elastic changes relaxation for layered structure 160 nm H_2Pc–20 nm Pd and 0.5% hydrogen in air at room temperature [28]

3.5 Theoretical Models of Coupling: Sensor Element-SAW

The theoretical models take into account the influence of these three main groups of parameters of the sensor elements: mechanical (density, modulus of elasticity), viscoelastic (viscosity), and electrical (electrical surface conductivity, dielectric constant) on the SAW propagation. This allows to analyze how a change in these parameters, as a result of an interaction with a gas, causes a disturbance of the SAW propagation. Changes in the attenuation and velocity of the SAW are the result of coupling of the parameters of the sensor element-SAW.

3.5.1 Types of Couplings

The realization of the sensor structure on the free surface, where the SAW propagates, introduces an initial load on the surface of the acoustic waveguide, which can then be further changed due to interaction with the gas molecules in its immediate environment. On the free surface, the SAW propagates with the speed v_0 and attenuation α, while after the sensor structure is created, there is a load on the waveguide, which causes a change in the speed of the wave propagation to the value v_0' (usually it is reduced), and an increase in the attenuation to the value α' (Fig. 3.12). This load on the acoustic waveguide surface results from the basic interactions (couplings) between the manufactured sensor element and the propagating SAW. These interactions in turn depend on the type of materials used (metallic, semiconductor, dielectric) and the geometric configuration (single, bilayer, multilayer structure) of the manufactured sensor element.

The most natural and most commonly used interaction is the *mass effect*, which consists in the mechanical loading of the waveguide surface by the manufactured sensor structure. The mechanical displacements are disturbed by the performance of the sensor element depending on its density, thickness, and elastic parameters. In addition, for polymer structures that are not perfectly elastic, their viscosity, hence the *viscoelastic effect*, will be important. The *acoustoelectric effect* is important for sensor elements made of semiconductors or metals, e.g. single semiconductor layers or bilayer semiconductor–metal structures.

In general, it should be assumed that the observed disturbances in the velocity and attenuation of SAW are the result of three types of interactions, the relative proportion of which depends on the mechanical and electrical properties of the sensor elements made. In addition, the small values of these disturbances and their dependence on the type of materials used allow each effect to be considered separately. This is the

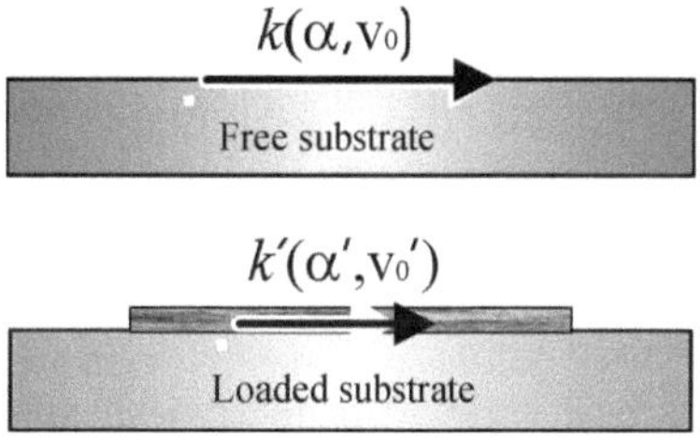

Fig. 3.12 SAW with a propagation constant k on a free substrate and k' on substrate loaded by the sensor element

principle of the additiveness of small disturbances, although these interactions are related to each other by the equations of the piezoelectric phenomenon [4].

The total effect of the SAW propagation perturbation, as a result of the execution of any sensor element (in the case of minor perturbations), will be approximately the sum of three component perturbations:

$$\frac{\Delta k}{k_0} \approx \left(\frac{\Delta k}{k_0}\right)_m + \left(\frac{\Delta k}{k_0}\right)_{ve} + \left(\frac{\Delta k}{k_0}\right)_{AE} \tag{3.10}$$

where $\Delta k = k' - k$ is the difference of the complex constant propagation k' for the wave disturbed by the performance of the sensor element and the constant propagation k for the free and undisturbed surface of the crystal on which the SAW propagates and k_0 is the wave number.

The index m means a disturbance arising from the mass effect, which means interactions related to both the surface density and the layer elastic constant or the entire sensor structure, *ve*—a disturbance arising from viscoelastic effect, while *AE*—acoustoelectric.

Mass (m) and viscoelastic (*ve*) interactions belong to the same group of mechanical disturbances, but for historical reasons they will be considered separately. A perfectly elastic, non-conductive, and non-piezoelectric sensor structure applied to a piezoelectric substrate affects SAW propagation and causes mass-elastic (*me*) disturbances. Such structures do not introduce additional wave attenuation. On the other hand, a sensor element with imperfect elasticity, simultaneously non-conductive and non-piezoelectric, causes viscoelastic interactions (*ve*), which cause additional wave attention.

The interactions between the made sensor elements and the acoustic waves used in gas sensors of various kinds are almost identical. The differences lie in the relative share of these couplings for a given sensor type. Depending on the type of piezoelectric substrate used (mainly the value of the electromechanical coefficient K^2), the configuration of the sensor element (single, double, or multiple structure) and the type of materials used (metallic, semiconductor, dielectric), the relative share of the mass and acoustoelectric effect in the SAW propagation disorder will be different.

For example, the acoustoelectric interaction will not be relevant for SAW sensors in which sensor elements with very high (metallic structures) and very low (dielectric structures) electrical conductivity are used. The use of metallic sensor structures on the surface of the piezoelectric crystal completely shields the influence of this effect. Thus, for STW-type sensors that use metallic strips on the surface, this effect will be significantly reduced. At the same time, for this type of sensor, mass interactions will be increased due to increased energy density. The mass effects of acoustic wave sensors are closely related to the distribution of acoustic wave energy density.

In addition to bulk acoustic wave (BAW) sensors (e.g. QCM) and acoustic plate mode (APM), for all other types of acoustic waves, this distribution is limited to surface areas with micrometer values. Thus, it is possible to achieve large mass impacts depending on the type of acoustic wave used and the sensor element made.

The direct effect of any type of interaction (mass, viscoelastic, or acoustoelectric) of the sensor element with SAW is the change of propagation velocity and attenuation. The magnitude of these disturbances will be determined within simple theoretical models, using the complex constant propagation.

3.5.2 Perturbation of Complex Propagation Constant

As a result of the production of a sensor element on the surface of the piezoelectric crystal, where SAW is propagating, there is a change in its constant propagation k, which in the general case has a complex form [59]:

$$k = \alpha + ik_0 \tag{3.11}$$

where α represents the attenuation of the wave, thus reducing its amplitude, while k_0 is the wave number $k_0 = \omega_0/v_0$ (describes the oscillation character of the wave), v_0 is the speed of SAW on free substrate (without sensor element).

The interaction of SAW with the sensor element leads to a disturbance of its complex propagation constant by the value Δk. The relative changes of this magnitude (relative to the undisturbed wave number k_0) are important, which for the assumed form (3.11) is

$$\frac{\Delta k}{k_0} = \frac{\Delta \alpha}{k_0} - i\frac{\Delta v}{v_0} \tag{3.12}$$

where relative changes in attenuation $\Delta\alpha/k_0$ represent the real part and relative changes in propagation velocity $\Delta v/v_0$ represent the imaginary part.

By means of appropriate theoretical methods, it is possible to find approximate expressions for the relative changes of the complex constant propagation, and thus to determine the relative changes of attenuation and SAW velocity. Depending on the properties of the sensor structures, made on the surface of the piezoelectric crystal, only their mechanical or electrical parameters can be taken into account. The most well-known method of determining the relative changes of constant propagation $\Delta k/k_0$ for each of the interactions discussed is the use of disturbance theory [4, 21, 53]. So-called coupled modes (e.g. [7]), normal modes [51], as well as an appropriately adapted transmission line method [85] can also be used for the acoustoelectric effect.

The perturbation theory is sufficiently accurate for the small values of the product of the wave number k_0 and the thickness of the sensor element h. For the sensor structures used in the study, this product did not exceed the value $k_0h \sim 2 \times 10^{-3}$, which justifies this type of analysis. The application of disturbance theory is based on the assumption that the manufactured sensor element represents a small disturbance of the mechanical and electrical boundary conditions relative to the free surface of the crystal.

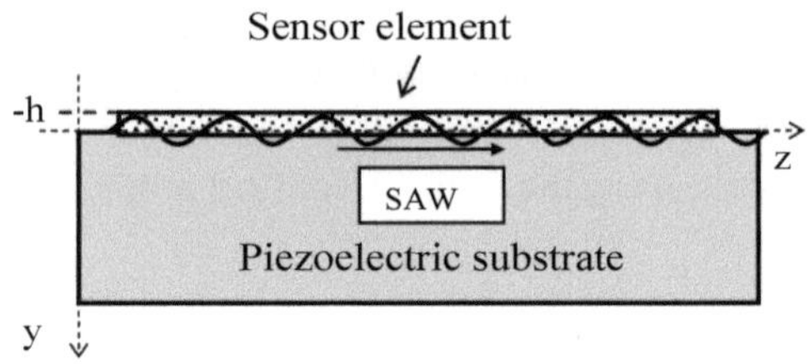

Fig. 3.13 Schematic representation of the system geometry: sensor element—piezoelectric substrate and system of coordinates

With respect to the specific geometry of the system (Fig. 3.13), perturbation theory can be used to determine a change (perturbation) of the complex constant propagation of the excited surface wave, where the source of the perturbation is a thin sensor element of h thickness generated directly on the surface of the crystal. The surface wave propagates along the z-axis in the plane XY for $y = 0$. Perturbations can only be considered in this plane ($y = 0$). Such approximation is acceptable as long as the thickness of the sensor element is significantly less than the wavelength of the propagating wave.

Using the perturbations theory, the change in the constant propagation of SAW as a result of the formation of a thin sensor element on the substrate surface has the form [52]:

$$\Delta k = \frac{-i\left\{-v_j^* T'_{yj} - v'_j T^*_{yj} + i\omega\left(\varphi' D^*_y - \varphi * D'_y\right)\right\}\Big]_{y=-h}^{y=0}}{\int_{-h}^{0} \left\{v_j^* T'_{zj} + v'_j T^*_{zj} + i\omega\left(\varphi\prime D^*_z - \varphi^* D'_z\right)\right\} dy} \tag{3.13}$$

where v_j—the velocity of points on the surface of the substrate, for which the dependence on time type $\exp(-i\omega t)$ is assumed, T_{ij}—the component tensor of mechanical stresses, φ—the electric potential, D_j—the component vectors of electrical induction, ′—the disturbed quantity, *—coupled.

Expression (3.13) requires an approximate definition of the disturbed fields $\boldsymbol{v}'$ and $\boldsymbol{T}'$, which is always a fundamental task for the application of perturbation theory [4]. Since the expected disturbances of the wave number due to the change of both mechanical and electrical boundary conditions are of the order of a few percent at most, and thus basically the disturbances are small, the disturbed fields in the denominator of Eq. (3.13) can be converted to undisturbed (the numerator is small and the denominator is large), giving [52]

$$4P = 2\mathrm{Re} \int_{-h}^{0} \left[-v_j^* T_z + i\omega\varphi D_z^*\right] dy \tag{3.14}$$

where P is the average wave power flux per unit of the acoustic beam width along the x-axis.

A similar simple approximation cannot be made in the expression numerator (3.13), because that would give $\Delta k = 0$. Therefore, another approximation, more accurate, must be found. The approximate perturbation formula for boundary perturbations is as follows [52]:

$$\Delta k \approx \frac{-i\left[-v_j^* T'_{yj} - v'_j T^*_{yj} + i\omega\left(\varphi' D^*_y - \varphi^* D'_y\right)\right]_{y=-h}^{y=0}}{4P} \tag{3.15}$$

Expression (3.15) correlates disturbed fields with undisturbed fields. To find an approximate change in the constant propagation of Δk, it is necessary to determine how undisturbed fields are related to disturbed fields. The first two components define mechanical disturbances, while the third and fourth components define electrical disturbances, resulting from the mechanical and electrical parameters of the made sensor structures, respectively.

3.5.3 Perturbations for Mechanical Effects

In this case, it is assumed that the sensor elements are non-conductive and non-piezoelectric. We do not consider the acoustoelectric interactions, which depend on their electrical parameters (mainly electrical conductivity). The disturbance of SAW propagation results from mechanical interactions that occur between the surface wave and the sensor coating. The sensor structure rigidly attached to the surface of the crystal undergoes both displacements (translations) and deformation due to the propagating wave. The translations essentially cause the so-called pure mass effect, i.e. a reduction in the speed of the wave propagation. On the other hand, the deformation of the structure causes energy to accumulate in it, which not only results in changing the speed of propagation of the wave, but also causes its attenuation as a result of the dispersion of energy in the structure.

When a thin sensor structure ($h \ll \lambda$) is produced on the surface of the substrate on which the SAW is propagated, the displacements on this surface (u_{x0}, u_{y0}, u_{z0}) do not change in the first approximation [4]. Propagation of the SAW results in the formation of a periodic displacement field, which is transmitted to the structure through its lower surface (contact) to the upper surface. Stresses are induced that depend on the mechanical properties of the materials from which the structure is made. Two cases can be distinguished [59]: (1) **acoustically thin** structures—where the u_i displacements are constant along the thickness of the structure, but only gradients of displacements are produced in its plane and (2) **acoustically thick** structures—where the displacements change not only in its plane, but also in a straighter direction.

As a result of the inertia of the structure, in particular of imperfectly elastic layers such as soft polymer layers, gradients in the displacement u_z in the direction perpendicular to the structure cause shear deformation. In general, thin and rigid structures behave like acoustically thin because the entire layer structure then moves synchronously with the substrate, so these displacements are uniform along its thickness. On the other hand, structures that are thick and soft behave like acoustically thick: the top surface lags behind the substrate displacement, resulting in non-uniform displacements along the sensor structure.

The parameter defining the boundaries between acoustically thin and thick structures is the number $R*$, given by the formula [59]:

$$R^* = \frac{A_p f v_0 \rho h}{|G|} \tag{3.16}$$

where A_p—constant; ρ, h, and $G = G' + iG''$ are the density, thickness, and shear modulus of the material of a single structure or their weighted mean with respect to ρ and G, and the sum of the thickness for multilayer systems; and f—frequency of oscillation.

The constant A_p depends on the parameters of the substrate material and is

$$A_p = 2\pi (c_x + c_z)^{1/2} \Big/ \left[(c_x + c_y)^{1/2} + 2c_z^{1/2} \right] \tag{3.17}$$

The values of the coefficients $c_j = v_{jo}{}^2/4k_0P$ are given in Table 3.3, for the most commonly used $LiNbO_3$ Y–Z substrate and ST-X quartz. The components v_{jo} denote the velocities of the contact surface of the substrate with the sensor element ($v_{jo} = i\omega u_{jo}$) in each of the directions $j = x, y, z$ and P is the SAW power per unit width. The estimated dimensionless constant A_p for $LiNbO_3$ Y–Z is about 3.48.

The c_j parameters are defined for the SAW propagation in the substrate without sensor coverage and it is assumed that they do not change considerably as a result of the formation of a thin layer or sensor structure.

The parameter $R*$ determines the proportion of stresses arising as a result of gradients of displacements in the direction perpendicular to the sensor structure relative to those in the plane of the structure.

When the sensor structure is sufficiently thin ($h \ll \lambda$) and rigid (large G) and at the same time the SAW frequency is such that $R* < 1$, then the displacement gradients in the plane of the structure dominate those along the transverse directions and it can be considered acoustically thin. On the other hand, for which $R* \geq 1$, inertia due to weak rigidity results in large displacement gradients along the thickness of the structure, it is then considered acoustically thick.

Table 3.3 Coupling parameters c_j between SAW and the substrate $LiNbO_3$ Y–Z and quartz ST-X

Displacement component	$c_j = v_{jo}{}^2/4k_0P \times 10^{-8}$, m²/s/kg	
	$LiNbO_3$ Y–Z	Quartz ST-X
u_x	0	0.013
u_y	2.403	1.421
u_z	1.101	0.615

Martin et al. [59] own estimation for $LiNbO_3$

The above delimitation, into acoustically thin and thick sensor structures, allows a different surface mechanical impedance to be determined for each category. The mechanical impedance is defined as [4]

$$Z_{mj} = \frac{T_{yj}}{v_j}\Big|_{y=0} \tag{3.18}$$

where T_{yj} and v_j determine the stress and velocity of the particle in the direction j at the boundary of the structure (for $y = 0$). The higher the mechanical impedance, the more difficult it is to deform the structure as a result of surface wave propagation. It depends on the mechanical properties of the structure and the type of deformation.

For acoustically thin structures ($R^* < 1$) the Z_{mj} impedance can be defined as [59]

$$Z_{mj} = i\omega h\left(\rho - \frac{E^{(j)}}{v_0^2}\right) \tag{3.19}$$

where ρ is the density of the single structure or weighted average for layered structures, h is the thickness of the entire structure, and v_0 is the undisturbed surface wave velocity, whereas

$$E^{(j)} = \left(\frac{1+\delta_{jz}}{2}\right)\frac{T_{jz}}{S_{jz}} \tag{3.20}$$

is a deformation modulus, taking into account the relative ratio of the stress T_{jz} to the deformation S_{jz}, in which $\delta_{jz} = 1$ for $j = z$ and 0 for $j = x, y$. The parameter $E^{(j)}$ is important for determining the contribution of the deformation to the surface wave disturbance. Three main types of this deformation can be distinguished: (1) *transverse shear* caused by the displacement gradient $\partial u_x/\partial z$, (2) *bending moment* as a result of $\partial u_y/\partial z$, and (3) *longitudinal compression (compression)* as a result of the $\partial u_z/\partial z$ gradient. The modulus values of $E^{(j)}$ for each type of deformation can be determined quite easily. They are given in Table 3.4.

Table 3.4 Deformation modules associated with the resulting deformation due to SAW propagation for acoustically thin layers ($R^* \ll 1$)

Type of deformation	Gradient of displacement	$E^{(j)}$ expressed in Lame constants λ' and μ' of the film	$E^{(j)}$ expressed by the spring modules K' i G' of the film
Transversal cutting	$\frac{\partial u_x}{\partial z}$	μ'	G'
Bending	$\frac{\partial u_y}{\partial z}$	~0	~0
Longitudinal compression	$\frac{\partial u_z}{\partial z}$	$\frac{4\mu'(\lambda'+\mu')}{\lambda'+2\mu'}$	$\frac{4G'(3K'+G')}{3K'+4G'}$

Martin et al. [59]

Using Expression (3.15), taking into account only the first two components of the numerator for the mechanical perturbation, one obtains

$$(\Delta k) = \frac{i\left\{v_j^*(0)Z_{mj}v_j(0)\right\}}{4P} \tag{3.21}$$

In Expression (3.21), it is assumed that the velocity field of acoustic particles on the surface $y = 0$ has not changed, i.e. $v_j'(0) = v_j(0)$ and $T_{yj}(0) = 0$, which means no stresses in the undisturbed state (without sensor structure), Z_{mj}—surface mechanical impedance.

In general, for mechanical disturbance, taking into account mass-viscoelastic interactions, relative changes in constant propagation (3.12), using perturbative expression (3.21), will be associated with the Z_{mj} impedance relationship [59]:

$$\frac{\Delta k}{k_0} = \frac{\Delta \alpha}{k_0} - i\frac{\Delta v}{v_0} = \sum_{j=x}^{z} c_j Z_{mj} \tag{3.22}$$

where $c_j = v_{jo}{}^2/4k_0P$ are the normalized velocities of points on the surface of the waveguide crystal (see Table 3.3).

On the basis of (3.19) and (3.22), the relative mass-viscoelastic (*mve*) disturbance of SAW constant propagation as a result of the production of an acoustically thin structure is

$$\left(\frac{\Delta k}{k_0}\right)_{mve} = \frac{\Delta \alpha}{k_0} - i\frac{\Delta v}{v_0} = i\omega h \sum_{j=x}^{z} c_j \left(\rho - \frac{E^{(j)}}{v_0^2}\right) \tag{3.23}$$

Based on Expression (3.23), using the c_j parameters from Table 3.3, and for the deformation modules in Table 3.4, relative changes in constant propagation and then relative changes in speed and damping of SAW can be determined.

From a physical point of view, it can be observed that each component of the displacement has contribution in two components of Eq. (3.23). One, proportional to the density ρ, is produced as a result of the kinetic energy associated with the displacements in the structure, while the other, proportional to $E^{(j)}$, is produced as a consequence of the potential energy associated with its deformations. The kinetic portion causes a decrease in the speed of the SAW proportional to product ρh (i.e. density per unit of surface area-surface density), which is referred to as the pure mass effect (pure mass load). The potential component, on the other hand, is proportional to the product $c_jE^{(j)}$, which causes changes in the propagation velocity of the SAW due to the deformation of the sensor structure.

3.5.3.1 Ideally Elastic Structures

For acoustically thin structures ($R^* \ll 1$), with perfectly elastic coverage, the moduli of deformation type $E^{(j)}$ have a real value (Table 3.4). This leads to pure mass-elastic interactions (*me*), since Expression (3.23) will only have imaginary value, which results in two equations:

$$\frac{\Delta\alpha}{k_0} = 0 \tag{3.24}$$

$$\left(\frac{\Delta v}{v_0}\right)_{me} = -\omega h\left[c_x\left(\rho - \frac{\mu'}{v_0^2}\right) + c_y\rho + c_z\left(\rho - \frac{4\mu'}{v_0^2}\frac{\lambda' + \mu'}{\lambda' + 2\mu'}\right)\right] \tag{3.25}$$

The ideal elastic coverage ($K' = G' = 0$) does not introduce damping. The expression (3.25) is known as Wohltjen's Equation [88], although the mechanical boundary conditions for this type of sensor structure were first defined by Tiersten and Shina [81].

For the $LiNbO_3$ Y–Z substrate frequently used in the studies the parameter $c_x = 0$, hence we get a simplified expression:

$$\left(\frac{\Delta v}{v_0}\right)_{me} = -\omega h\left[c_y\rho + c_z\left(\rho - \frac{4\mu'}{v_0^2}\frac{\lambda' + \mu'}{\lambda' + 2\mu'}\right)\right] \tag{3.26}$$

which can also be presented in another form [32]:

$$\left(\frac{\Delta v}{v_o}\right)_{me} \approx \underbrace{-c_{m_1} f_o \rho_{srf}}_{\mathbf{S_1}} + \underbrace{c_{m_2} f_o h \mu' \frac{\lambda' + \mu'}{\lambda' + 2\mu'}}_{\mathbf{S_2}} \tag{3.27}$$

where c_{m_1}—mechanical coupling constant dependent on substrate parameters of ~5.5×10^{-8} m^2s/kg for $LiNbO_3$ Y–Z ("pure mass effect"), $\rho_{srf} = \rho\ h$—average surface density of the structure, h—total thickness of the structure, ρ—average surface density, c_{m_2}—"stiffness" constant of ~5.7×10^{-15} [s^3/kg] for $LiNbO_3$ Y–Z; λ', μ'—Lame' constant, average values for the structure, and f_0—wave frequency on the free surface of the crystal.

On the basis of Expression (3.27), it can be seen that the formation of a thin isotropic sensor structure on the surface of the piezoelectric, depending on its mechanical properties, results in a "pure" mass load (the first negative component $\mathbf{S_1}$) or a "stiffening" (the second positive component $\mathbf{S_2}$) of the waveguide surface. As a result, depending on which of the components of Eq. (3.27) prevails, we get a decrease or increase in the speed of the surface wave propagation. The component of the pure mass load $\mathbf{S_1}$ is dependent on the wave frequency f_0 and on the surface density of the sensor structure, i.e. the mean volume density and the total thickness h.

On the other hand, the component of stiffening of the surface $\mathbf{S_2}$, causing the increase of the wave velocity, depends on the mean parameters of the elastic structure μ' and λ', as well as on the total thickness h and the frequency of propagation f_0.

In many cases, the omission of the second component in (3.27) is justified by its small contribution to the total change in SAW velocity. An even simpler expression is then obtained taking into account only the pure mass effect [32]:

$$\left(\frac{\Delta v}{v_o}\right)_m \approx -c_{m_1} f_o \rho_{srf} \tag{3.28}$$

Expression (3.28), after taking into account the coverage factor κ (the length of the sensor structure in relation to the entire length of the delay line) and the fact that ρ_s in the case of the sensor response will be a change in this density as a product of the number of adsorbed molecules per unit of surface area N_s and their individual masses m_1, gives the equation [60]:

$$\frac{\Delta f}{f_0} = \kappa \frac{\Delta v}{v_0} = -\kappa c_{m1} f_0 N_s m_1 \tag{3.29}$$

where the product of constant is c_{m1} and frequency f_0 is the mass sensitivity S_m, defined by formula (2.2). Relative frequency changes are described only by the pure mass effect, which is roughly true for structurally stabilized sensor elements, e.g. phthalocyanines, metal oxides, or single metal layers. Acoustoelectric couplings do not occur.

3.5.3.2 Viscoelastic Structures

For viscous and acoustically thin coatings ($R^* \ll 1$), complex expressions for the structural elasticity coefficients $K = K' + K''$, $G = G' + G''$ should be applied for the deformation modules $E^{(j)}$ in (3.23). A simplification here will be to take into account the fact that the volumetric moduli for soft, flexible materials are usually much larger than the shear moduli ($K \gg G$), leading to $E^{(z)} \sim 4G$. The relative changes in damping (viscoelastic) and velocity (caused by mass-elastic interactions) for such a structure are [59]

$$\left(\frac{\Delta \alpha}{k_0}\right)_{vel} \cong \frac{\omega h}{v_0^2}(c_x + 4c_z)G'' \tag{3.30}$$

$$\left(\frac{\Delta v}{v_0}\right)_{mel} \cong -\omega h\left[(c_x + c_y + c_z)\rho - \frac{(c_x + 4c_z)}{v_0^2}G'\right] \tag{3.31}$$

where the factors c_x, c_y, c_z are found in Table 3.3, while $G'' = \omega\eta$ is the product of frequency and dynamic viscosity. In addition to changes in propagation velocity, viscoelastic structures cause additional changes in SAW attenuation.

3.5.3.3 Acoustically Thick Structures

If the parameter $R^* \geq 1$, then the following expression can be obtained for the relative change in constant SAW propagation for mass-viscoelastic perturbation [59]:

$$\left(\frac{\Delta k}{k_0}\right)_{mve} = \frac{\Delta \alpha}{k_0} - i\frac{\Delta v}{v_0} = \sum_{j=x}^{z} \frac{c_j \beta_j M_j}{\omega} \tanh\left(i\beta_j h\right) \tag{3.32}$$

where β_j parameters represent the propagation constants for the particle waves generated into the structure and $M_x = M_z = G$ and $M_y = K$. Resultant displacements are the superposition of waves generated at the substrate-structure interface as a result of displacements on the substrate surface u_{j0}, which then propagate to the structure with constant propagations β_j. The normal component to the surface u_{y0} generates a compressional wave, while the components in the substrate planes u_{x0} and u_{z0} generate shear waves. These constants can be expressed as

$$\beta_j = \omega \left(\frac{\rho - E^{(j)}/v_0^2}{M_j} \right)^{1/2} \tag{3.33}$$

For acoustically thin covers, Expression (3.32) is reduced to (3.24) and (3.25) because $\tanh(i\beta_j h) \sim i\beta_j h$ dla $\beta_j h \ll 1$. The deformation modules $E^{(j)}$ are shown in Table 3.4.

Acoustically thick structures will be important for higher values of $\beta_j h$, i.e. for structures with large β_j or significant h-thicknesses. In turn, constant β_j will be important for structures with high densities and small $E^{(j)}$ and small M_j moduli and high wave propagation frequencies.

3.5.3.4 Parameter *R**

The value of the parameter R^* defined by formula (3.16) and distinguishing between acoustically thin and thick structures can be estimated for the used sensor coverings on the basis of the available material parameters and data for the most commonly used piezoelectric substrate type $LiNbO_3$ Y–Z.

The thin layers of CuPc phthalocyanine obtained by vacuum evaporation are characterized by elastic moduli in the ranges: $\mu' = G' = \sim 0.5$–5 GPa, Young's modulus of E_{CuPc} ~9.3 GPa [48], and CuPc density of ~1630 kg/m^3. On the other hand, thin layers of palladium (10–100 nm) have a Young's module of $E_{Pd} = 115$ GPa (about 7% lower than the bulk structures of ~124 GPa) [83]. The elastic moduli for nanometer palladium layers are more than an order of magnitude lower than Young's module. The density of palladium is $\rho \sim 12{,}023$ kg/m^3.

Estimation of the parameter R^* for both phthalocyanine and palladium thin layers produced on $LiNbO_3$ Y–Z substrates with parameters $v_0 \sim 3500$ m/s and $A_p \sim 3.48$ and wave frequency $f \sim 43$ MHz gives the following results:

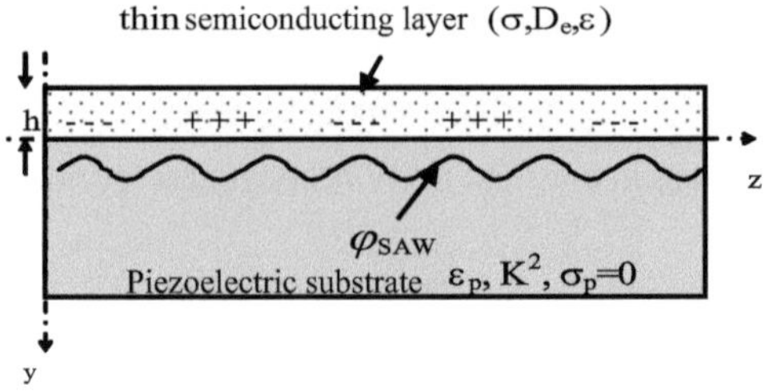

Fig. 3.14 Electric potential of the SAW excited on piezoelectric substrate interacts with charge carriers in a thin semiconductor film, causing their separation and wave perturbation as a result

h_{CuPc} ~ 80 – 720 nm $\boldsymbol{R}^*$ ~ **0.070 – 0.600** acoustically thin structures,

h_{Pd} ~ 10–30 nm $\boldsymbol{R}^*$ ~ **0.006 – 0.020** also acoustically thin.

The parameters for the MPc-Pd layered structures will be between these values, so they can be considered acoustically thin. The use of Expressions (3.24)–(3.29) to estimate the mass effects perturbing the SAW propagation is therefore justified for this type of sensor elements.

3.5.4 *Perturbations for Acoustoelectric Effect*

In the case of semiconductor sensor structures, which are characterized by appropriate electrical parameters, i.e. electrical conductivity, charge carrier constant diffusion, and electrical permeability, the SAW propagation is additionally disturbed by the acoustoelectric effect. The electrical potential φ associated with the surface wave propagating in piezoelectric can interact with the electric charge carriers located in the layer or sensor structure (Fig. 3.14). This coupling, depending on the layer or structure parameters, results usually in a decrease in the propagation velocity and an increase in the SAW attenuation.

3.5.4.1 Single Structures

For semiconductor sensor elements in the form of single thin layers (Fig. 3.14) with a thickness $h \ll \lambda$, it is assumed that the mechanical boundary conditions are undisturbed ($T_{yj} = T_{yj}' = 0$). The perturbation formula (3.15), for the acoustoelectric effect, gets the following form in the plane XZ for $y = 0$ [52]:

$$(\Delta k)_\sigma \approx \frac{\omega\left[\varphi'(0)D_y^*(0) - \varphi^*(0)D_y'(0)\right]}{4P} \tag{3.34}$$

The best agreement between disturbance theory and precise numerical calculations for single layers is obtained in the so-called weak coupling approximation, in which the invariability of the stress field $\boldsymbol{T}$ in the perturbation calculus is assumed [52, 81].

Using the perturbation formula in the form of (3.34), it is possible to obtain disturbances of the electrical boundary conditions resulting from the formation of a thin sensor layer in the plane $y = 0$. The relative change of the SAW constant propagation is then [4]:

$$\left(\frac{\Delta k}{k_0}\right)_{AE} = \frac{K^2}{2} \frac{1 + iz_E'(0)}{1 - i\frac{\varepsilon_p^T}{\varepsilon_0} z_E'(0)} \tag{3.35}$$

where z_E' is the normalized surface electrical impedance of the layer, defined in the plane $y = 0$. This impedance, for a single layer, is defined as the ratio of the disturbed electrical potential to the disturbed component electrical induction for $y = 0$ [4]:

$$z_E'(0) = -ik_0\varepsilon_0 \left(\frac{\varphi'}{D_y'}\right)_{y=0} \tag{3.36}$$

whereas

$$\varepsilon_p^T = \sqrt{\varepsilon_{yy}^T \varepsilon_{zz}^T - \left(\varepsilon_{yz}^T\right)^2} \tag{3.37}$$

is the effective electrical permeability of the piezoelectric substrate, with a constant stress field $\boldsymbol{T}$. To determine the z_E' impedance, the equations describing the motion of electric charge carriers in a semiconductor should be used [53]:

(a) Equation of current density:

$$J = \mu\rho_e \mathrm{E} - D_e \nabla \rho_e \tag{3.38}$$

(b) Equation of continuity:

$$\frac{\partial \rho_e}{\partial t} + \nabla J = 0 \tag{3.39}$$

(c) Electrostatic approximation:

$$E = -\nabla\varphi \tag{3.40}$$

where μ is the mobility of electric carriers, ρ_e is the spatial density of electric charges, D_e is their constant diffusion, E is the strength of the electric field, and φ is the electric potential.

Furthermore, it should be assumed that the thickness of the layer $h \ll \lambda$ and $h \ll \lambda_D$. The Debye shielding length, $\lambda_D = (\varepsilon D_e/\sigma)^{1/2}$, characterizes the depth of penetration of the electric field into the space above the surface of the piezoelectric crystal. This allows for the normalized electrical surface impedances of the single structure to be obtained. In the simplest case of considering only the surface conductivity of the layer $\boldsymbol{\sigma_s = \sigma h}$, one obtains [45]

$$z'_E(0) = \left(\frac{\sigma_s}{\varepsilon_0 v_0} - i\right)^{-1} \tag{3.41}$$

The electrical surface impedance (3.41) is a function of only the surface conductivity of the layer when the diffusion effects are small ($D_e \sim 0$).

The constant of propagation in the general case is the complex quantity. For a form (3.11), the relative change in the speed of propagation of a wave will be equal to the relative change of its imaginary part with a minus sign, while the change in attenuation relative to the undisturbed wave number—its real part. On the basis of (3.35) and (3.41), we get [45, 68]

$$\frac{\Delta\alpha}{k_0} = \mathrm{Re}\left\{\frac{\Delta k}{k_0}\right\} \approx \frac{K^2}{2}\frac{v_0 C_s \sigma_s}{\sigma_s^2 + v_0^2 C_s^2} \tag{3.42}$$

$$\frac{\Delta v}{v_0} = -\mathrm{Im}\left\{\frac{\Delta k}{k_0}\right\} \approx -\frac{K^2}{2}\frac{\sigma_s^2}{\sigma_s^2 + v_0^2 C_s^2} \tag{3.43}$$

where K^2 is the electromechanical coefficient, for $LiNbO_3$ Y–Z it is 0.048; $\Delta v = v - v_0$ is the change in the propagation velocity of the disturbed wave v, in relation to the undisturbed wave velocity $v_0 = 3487.7$ m/s on the free surface of the crystal $LiNbO_3$ Y–Z; $C_s = \varepsilon_0 + \varepsilon_p$ is the sum of the absolute electrical permeability of the environment and the substrate and is 4.6 pF/cm for air and $LiNbO_3$ Y–Z; and $\sigma_s = \sigma h$ is the surface conductivity of the thin sensor layer.

The graph of relative changes in propagation velocity and acoustic attenuation of SAW, as a function of the normalized surface conductivity of the thin film, is shown in Fig. 3.15. The largest changes in velocity and greatest attenuation for the surface wave occur when $\boldsymbol{\sigma_s = v_0 C_s = 1.6 \times 10^{-6}\ \Omega^{-1}}$, in the case of the $LiNbO_3$ Y–Z substrate most commonly used in the studies.

The propagation velocity of SAW decreases as a result of the formation of a thin semiconductor layer. This decrease in velocity is the result of the acoustoelectric interaction of the electric potential associated with the surface wave and the moving charge carriers in the layer. As a result of this interaction, energy is transferred from the wave to the carriers, resulting in a disturbance in the propagation speed and an increase in acoustoelectric attenuation. Details of this approach can be found in [68] and [14]. It is worth noting that with the exception of diffusion effects, e.g. for phthalocyanines with very low constant diffusion of charge carriers [50], the disturbance of wave propagation does not depend on the frequency of oscillation, which is important in sensor applications. Gas sensors, which use the acoustoelectric effect to achieve high sensitivity, do not require high SAW frequencies.

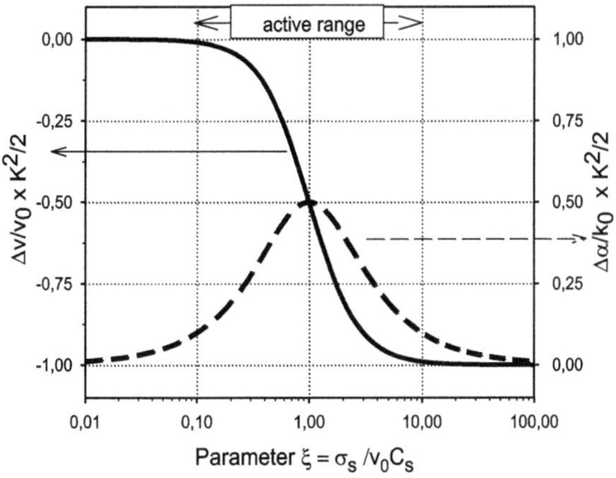

Fig. 3.15 Relative changes of the acoustoelectric attenuation $\Delta\alpha/k_0$ and propagation velocity of the SAW $\Delta v/v_0$ as a function of the normalized film electric surface conductivity σ_s placed on piezoelectric substrate with electromechanical coefficient K^2

On the basis of Fig. 3.15, it can also be seen that the active acoustoelectric interaction range is limited by the parameter ξ, i.e. the ratio of the conductivity of the surface layer σ_s to the product $v_0 C_s$, which in turn characterizes the properties of the piezoelectric substrate used. The active acoustoelectrical interaction range is the range of the parameter ξ in which this interaction is particularly strong. This range includes approximately four orders of magnitude:

$$0.01 < \xi < 100 \tag{3.44}$$

which for the most commonly used substrate $LiNbO_3$ Y–Z gives

$$1.6 \times 10^{-8} \Omega^{-1} < \sigma_s < 1.6 \times 10^{-4} \Omega^{-1} \tag{3.45}$$

For single semiconductor sensor structures obtained by the author of this work, the electrical surface conductivity values are in the vicinity of the end of the lower range (3.45), which is mainly due to the low value of their thickness. For organic semiconductors (metalophthalocyanine), the value of σ_s, measured by the author by direct electrical method, is in this range, however, after taking into account the strong dependence of this value on the frequency of $\sigma_s(f \sim 43 \text{ MHz}) \sim 10^{-7}\ \Omega^{-1}$ [45, 69]. Metal layers are characterized by much better electrical conductivity, therefore their σ_s values generally exceed the given upper range of (3.45). At very small thicknesses of metal layers (~1–3 nm), however, their conductivity decreases, which results from the limitation of the mean free path of electric charge carriers [19].

One way to better match semiconductor sensor structures to the active range of acoustoelectric interactions is to increase their temperature. Thermal activation causes an increase in the electrical conductivity σ_s of most semiconductor layers, which in the case of structures characterized by acoustoelectric parameters $\xi =$ 0.01–0.1 allows to increase their value to a small extent. A much larger change in the incident conductivity can be achieved structurally, by making a layered structure, e.g.

a semiconductor–metal. Placing a thin metal layer on a semiconductor layer increases the incident conductivity σ_s of the entire structure by up to 5–6 orders of size. The acoustoelectric parameter of the entire layer structure then undergoes a significant shift towards higher values, increasing the possibility of obtaining strong interactions. Layered structures are a new type of sensor element in SAW sensors, through greatly increased possibilities of using acoustoelectric interactions [37, 35, 34].

3.5.4.2 Layered Structures

For bilayer structures, the influence of the interrelated surface charges induced in each layer by the SAW-accociated electrical potential, as well as the finite electrical conductivity of each layer, must be taken into account. The parameters of the layer structure that significantly affect the disturbance of SAW propagation are the electrical conductivity of the layers σ_1 and σ_2 and their thicknesses h_1 and $h_2 \ll \lambda$ (Fig. 3.16a). For simplicity, the influence of constant D_e diffusion and electrical permeability will be omitted.

The presented method of determining the SAW propagation disturbance is a modification of the energetic conception of the phenomenon proposed by Ricco et al. [68], Jakubik [29]. The issue has been generalized to bilayer structures. As before, the

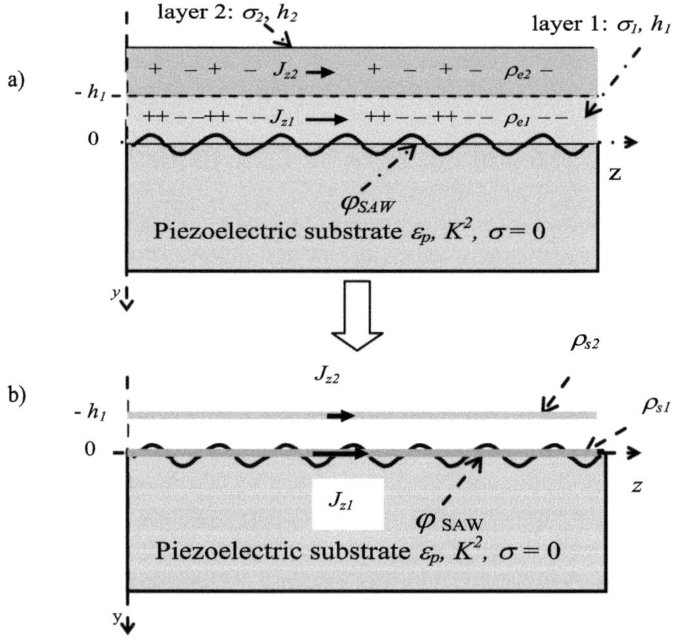

Fig. 3.16 **a** Electric boundary condition perturbation of the SAW by bilayer sensor structure characterized by electrical conductivities σ_1, σ_2 and thicknesses h_1 and h_2, **b** model using a surface density of the induced charges ρ_{s1} and ρ_{s2}

perturbation method will be used, but without the use of surface impedance, which is difficult to calculate due to the interrelationship of charge density in both layers [43, 44].

A bilayer structure, in which there is no spatial distribution of electrical charges ρ_e without SAW, is characterized by two substantially different electrical conductivities. Whereas the SAW excitation, through the associated electrical potential $\varphi_{AFP} = \varphi$, will result in two different density charges ρ_{e1} and ρ_{e2}, in each layer. The problem can be divided into two parts:

I. *Power transfer, via the electrical potential φ associated with SAW, to mobile charge carriers in both layers.*
II. *The effect of the transfer of power from SAW to its constant propagation.*

The determination of the power flux in a complex form, transmitted from SAW to charge carriers, makes it possible to determine the attenuation and change the speed of the wave propagation. The assumption that the thicknesses of the two layers are much smaller than the wavelength makes it possible to treat the issue in a one-dimensional manner (Fig. 3.16b).

I. *Power transfer from SAW potential φ to charge carriers in both layers*

The electrical potential associated with SAW is given as a formula:

$$\varphi(z, t) = \varphi_0 e^{i(\omega_0 t - k_0 z)} \tag{3.46}$$

where ω_0 and k_0 are undisturbed values of the frequency and wave number of the SAW.

When the conductive bilayer structure is placed on the crystal surface ($y = 0$) and in a closed space above it ($h_1 + h_2 \ll \lambda$), then the SAW propagation will undergo a slight disturbance:

$$\varphi_0(z) = \varphi_0(0)e^{-(\Delta k)z} \tag{3.47}$$

where $\Delta k = k' - k$ is the difference of the complex constant propagation of the disturbed wave $k' = \alpha' + ik_0'$ (by the structure manufacturing) and the complex constant propagation for the undisturbed wave $k = \alpha + ik_0$ (on the free surface of the piezoelectric crystal). It can be said that $\Delta k = \Delta\alpha + i\Delta k_0$ is a perturbation of the complex constant propagation, resulting from the interaction of the wave with the charge carriers in the bilayer structure. Since this interaction is usually small, it can be assumed that $|\Delta k| \ll k_0$.

Based on perturbation theory, small changes Δk in the complex constant propagation of SAW can be calculated on the basis of the power transferred to the moving charge carriers of each layer, by the undisturbed potential of the surface wave. As a result, the potential of φ_0 in (3.46) will be considered constant when calculating the power transfer from the wave to the moving charge carriers, i.e. only when using the pattern (3.47), i.e. when considering a change in the complex propagation constant. Assuming a small layer thickness, the spatial charge distribution can be replaced by the surface distribution. In the model shown in Fig. 3.16b, layer 1 with a charge density ρ_{e1} is replaced by a surface charge $\rho_{s1} = \rho_{e1}h_1$, placed in the plane $y = 0$, while layer 2, where the ρ_{e2} is present, is replaced by a charge $\rho_{s2} = \rho_{e2}h_2$ in the plane of $y = -h_1$.

The electrical surface potential φ_{AFP} induces the surface charges ρ_{s1} and ρ_{s2} of each layer. Current densities J_{z1} i J_{z2} are related to charge densities through continuity equations (without diffusion effects):

$$\frac{\partial}{\partial t}\left(\frac{\rho_{s1}}{h_1}\right) + \frac{\partial J_{z1}}{\partial z} = 0 \tag{3.48}$$

$$\frac{\partial}{\partial t}\left(\frac{\rho_{s2}}{h_2}\right) + \frac{\partial J_{z2}}{\partial z} = 0 \tag{3.49}$$

From these equations and (3.38)–(3.40) and (3.46), one obtains

$$\rho_{s1} = \frac{k_0 h_1}{\omega_0} J_{z1} \quad \text{i} \quad \rho_{s2} = \frac{k_0 h_2}{\omega_0} J_{z2} \tag{3.50}$$

The induced surface charges create an additional surface potential φ_{d1} from ρ_{s1} and φ_{d2} from ρ_{s2}, which together with the associated SAW potential φ creates the total surface potential $\varphi_t = \varphi + \varphi_{d1} + \varphi_{d2}$. The current densities can be determined by this total potential (designated for the corresponding plane) and the conductivity of each layer σ_1 and σ_2. Because $E_z = -\partial\varphi_t / \partial z$, so

$$J_{z1} = \sigma_1 E_{z1} = -\sigma_1 \frac{\partial}{\partial z}(\varphi + \varphi_{d1} + \varphi_{d2})_{y=0} = ik_0\sigma_1(\varphi_c)_{y=0} \tag{3.51}$$

In (3.51) φ_{d2} comes from the charge ρ_{s2}, but is calculated on the surface of $y = 0$. On the other hand, E_{z1} is the electric field strength. The diffusion current has been neglected, which is true when the SAW potential $\varphi \gg k_B T/q$. This condition is met on the surface of $LiNbO_3$ when the acoustic power density exceeds 0.2 mW/cm of the acoustic beam width [68].

The current density in the plane $y = -h_1$ will be similarly:

$$J_{z2} = \sigma_2 E_{z2} = -\sigma_2 \frac{\partial}{\partial z}(\varphi + \varphi_{d1} + \varphi_{d2})_{y=-h_1} = ik_0\sigma_2(\varphi_c)_{y=-h_1} \tag{3.52}$$

In (3.52) ϕ_{d1} comes from ρ_{s1}, but on the surface $y = -h_1$.

The additional surface potentials generated by the induced surface charges depend on the dielectric parameters of the substrate and the SAW wave number and are, respectively:

$$\varphi_{d1} = \frac{\rho_{s1}}{k_0 C_{s1}} \quad \text{and} \quad \varphi_{d2} = \frac{\rho_{s2}}{k_0 C_{s2}} e^{-k_0 h_1} \tag{3.53}$$

where $k_0 C_{s1}$ and $k_0 C_{s2}$ are capacitances per unit area between charges induced on the surface $y = 0$ and $y = -h_1$, respectively, by the SAW of the wave number k_0. It can be assumed that $C_{s1} \approx C_{s2} = C_s = \varepsilon_0 + \varepsilon_p$. The factor $\exp(-k_o h_1)$ in terms for ϕ_{d2} means the reduction of the potential as a result of the distance h_1 from the surface of the SAW propagation. For small values $k_0 h_1$, it does not matter much, but can be important for larger thicknesses of layer 1 and larger values of the wave number k_0, i.e. smaller wavelengths, which occur at higher frequencies of excited surface waves.

From Expressions (3.50) to (3.53), one can get the charges induced in each of the layers by the SAW potential. For each plane in which there is a surface charge replacing the actual layer, one obtains the corresponding dependencies $\rho_{s1} = f(\varphi, \rho_{s2})$ for $y = 0$ and $\rho_{s2} = f(\varphi, \rho_{s1})$ for $y = -h_1$. We get a set of equations, the solutions of which are the dependencies:

$$\rho_{s1} = \frac{k_0 \sigma_{s1}\left(i - \frac{\sigma_{s2}}{v_0 C_s} e^{-2k_0 h_1}\right)}{v_0\left[1 - i\left(\frac{\sigma_{s1}+\sigma_{s2}}{v_0 C_s}\right) + \frac{\sigma_{s1}\sigma_{s2}}{v_0^2 C_s^2}\left(1 - e^{-2k_0 h_1}\right)\right]}\varphi \tag{3.54}$$

$$\rho_{s2} = \frac{i k_0 C_s \sigma_{s2}}{v_0 C_s - i\sigma_{s2}}\left\{1 + \frac{\sigma_{s1}}{v_0 C_s}\frac{i - \frac{\sigma_{s2}}{v_0 C_s} e^{-2k_0 h_1}}{\left[1 - i\left(\frac{\sigma_{s1}+\sigma_{s2}}{v_0 C_s}\right) + \frac{\sigma_{s1}\sigma_{s2}}{v_0^2 C_s^2}\left(1 - e^{-2k_0 h_1}\right)\right]}\right\}\varphi e^{-k_0 h_1} \tag{3.55}$$

where $\sigma_{s1} = \sigma_1 h_1$ and $\sigma_{s2} = \sigma_2 h_2$, are the surface conductivities for the layers 1 and 2.

Expressions (3.54) and (3.55) represent the induced surface charges in the corresponding plane, replacing each of the layers by the potential φ of SAW. The parameters of the layers are their adequate electrical surface conductivities. The piezoelectric substrate, on the other hand, is characterized by the wave velocity v_0 and the parameter $C_s \approx \varepsilon_0 + \varepsilon_p$, while omitting the electrical permeability of each layer.

The movement of charges in both layers, as a result of the interaction with the SAW potential, is caused by the flow of power from the wave to the carriers of electrical charges. The principle of conservation of energy can be written as

$$P_y = P_{y1} + P_{y2} \tag{3.56}$$

$$P_{y1} = -\frac{\partial \rho_{s1}}{\partial t}\frac{\varphi(0)}{2} = -\frac{i\omega_0 \rho_{s1}\varphi(0)}{2} \tag{3.57}$$

$$P_{y2} = -\frac{\partial \rho_{s2}}{\partial t}\frac{\varphi(-h_1)}{2} = -\frac{i\omega_0 \rho_{s2}\varphi(-h_1)}{2} \tag{3.58}$$

where (3.57) and (3.58) describe the average power flux per unit area in the y-direction, perpendicular to the wave propagation direction, from SAW to the charge carriers, in layers 1 and 2, respectively.

From Eqs. (3.54)–(3.58), the total complex power flux from the wave to the charge carriers in both layers is

$$P_y = -\frac{i\omega_0}{2}[\rho_{s1}\varphi(0) + \rho_{s2}\varphi(-h_1)] \tag{3.59}$$

$$\varphi(-h_1) = \varphi(0)e^{-k_0 h_1} \tag{3.60}$$

Expression (3.59) is true, provided that $\lambda_D > h_1 + h_2$, which ensures that the SAW field penetrates each layer.

II. *SAW propagation disturbance due to transfer of power flux to charge carriers*

Power transfer to moving charge carriers in both layers comes from SAW. As a result, it perturbs its propagation, i.e. its speed and damping.

One can define the impedance Z_0, which links the surface potential φ *of* SAW on the surface $y = 0$ with the total wave power P_z [51]:

$$Z_0 = \frac{\varphi^2(0)}{2P_z} \tag{3.61}$$

On the other hand, with the propagating SAW is associated the energy density of u_T, representing the total (mechanical and electrical) energy accumulated per unit area. This energy density can be associated with the total average power of P_z SAW by the beam width a by

$$u_T = \frac{P_z}{av_0} = \frac{\varphi^2(0)}{2Z_0 av_0} \tag{3.62}$$

The total power flux transmitted by SAW to the charge carriers in both layers is given by changes in the energy density of u_T:

$$P_y = -v_0\frac{\partial u_T}{\partial z} = -\frac{1}{a}\frac{\partial P_z}{\partial z} \tag{3.63}$$

Equations (3.47), (3.61), and (3.63) show that the power transmitted by the wave is related to the change in the propagation of the SAW, by disturbing its constant propagation by Δk:

$$P_y = \frac{\Delta k \varphi^2(0)}{aZ_0} \tag{3.64}$$

In turn, the impedance Z_0 can be related to the electromechanical coupling coefficient K^2, which is used to characterize piezoelectric substrates. It is defined as the ratio of energy density in electrical form to total energy density:

$$K^2 = \frac{u_{E1} + u_{E2}}{u_T} \tag{3.65}$$

where

$$u_{E1} = \frac{1}{2} k_0 C_s \varphi^2(0) \text{ and } u_{E2} = \frac{1}{2} k_0 C_s \varphi^2(-h_1) \tag{3.66}$$

are the electric densities in layer 1 and layer 2, respectively, and the impedance can be determined as

$$Z_0 = \frac{K^2}{a v_0 k_0 C_s (1 + e^{-2k_0 h_1})} \tag{3.67}$$

The power transmitted by SAW to the charge carriers in both layers can be expressed as

$$P_y = \frac{(\Delta k) k_0 v_0 C_s \left(1 + e^{-2k_0 h_1}\right) \varphi^2(0)}{K^2} \tag{3.68}$$

Since the SAW propagation constant is expressed in the form (3.11) and its relative changes are described by Expression (3.12), the complex power transmitted by the SAW to the charge carriers (3.68) can be equated to the complex power received from the wave by the charge carriers (3.59) and (3.60). Knowing that $v_0 = \omega_0 / k_0$ and separating the real and imagined parts with respect to the wave number k_0 with limitation to second-order expressions, we obtain the relative changes in the attenuation and speed of the SAW as a result of producing a bilayer sensor structure on a piezoelectric substrate with a coupling coefficient K^2 [29]:

$$\frac{\Delta \alpha}{k_0} \approx \frac{K^2}{2\left(1 + e^{-2k_0 h_1}\right)} v_0 C_s \left[\frac{v_0^2 C_s^2 \left(\sigma_{s1} + \sigma_{s2} e^{-2k_0 h_1}\right) + \sigma_{s1} \sigma_{s2} \left(\sigma_{s2} + \sigma_{s1} e^{-2k_0 h_1}\right)}{v_0^2 C_s^2 \left(\sigma_{s1}^2 + \sigma_{s2}^2 + v_0^2 C_s^2\right) + \sigma_{s1}^2 \sigma_{s2}^2} \right] \tag{3.69}$$

$$\frac{\Delta v}{v_0} \approx -\frac{K^2}{2\left(1 + e^{-2k_0 h_1}\right)} \left[\frac{v_0^2 C_s^2 \left(\sigma_{s1}^2 + \sigma_{s2}^2 e^{-2k_0 h_1}\right) + \sigma_{s1}^2 \sigma_{s2}^2 \left(1 + e^{-2k_0 h_1}\right)}{v_0^2 C_s^2 \left(\sigma_{s1}^2 + \sigma_{s2}^2 + v_0^2 C_s^2\right) + \sigma_{s1}^2 \sigma_{s2}^2} \right] \tag{3.70}$$

To check the correctness of the above expressions, the limit values are applied. Assuming in the formulas (3.69) and (3.70) that $h_1 \to \infty$ and $\sigma_{s2} \to 0$, identical patterns are obtained as for a single sensor structure (3.42) and (3.43). The conditions of $h_1 \to \infty$ and $\sigma_{s2} \to 0$ are equivalent to the disappearance of the second layer, by moving the ideal dielectric layer indefinitely away from the first layer. An additional check is the limit values for the ideally conductive bilayer structure: σ_{s1} and $\sigma_{s2} \to \infty$ and $h_1 \to 0$. The values consistent with the ideally conductive single layer are then obtained: $\Delta\alpha/k_0 \to 0$ and $\Delta v/v_0 \to -K^2/2$.

Expressions (3.69) and (3.70) obtained by the author roughly describe the disturbances of SAW propagation as a result of the creation of a bilayer sensor structure, whose important parameters are its surface electrical conductivity $\sigma_{s1} = \sigma_1 h_1$ and $\sigma_{s2} = \sigma_2 h_2$. Compared to single layers, additional components appear with parameter σ_{s2}, which show the effect on the SAW propagation from the second layer. Additional parameters are the wave number k_0 (which depends on the wavelength λ) and the height h_1 above the piezoelectric surface, where the second layer is located. The dependence on k_0 means that even if the diffusion effects are omitted, bilayer structures present the dispersion effects. For single structures, this frequency dependence is visible only when the constant diffusion of the layer is taken into account [45].

Analysis in function of normalized conductivity ξ_1 of the first layer [29]

The results of the numerical analysis of relative velocity changes (formula 3.70) for the most commonly used values $k_0 = 7.85 \times 10^4\ \text{m}^{-1}$ (at $\lambda = 80\ \mu\text{m}$) and $h_1 = 250$ nm, as a function of the normalized surface conductivity of the first layer (parameter $\xi_1 = \sigma_{s1}/v_o C_s$) and the seven conductivity values σ_{s2} relative to σ_{s1} ($\sigma_{s2} = x\,\sigma_{s1}$, where $x = 0, 0.001, 0.01, 5, 10, 100$, and 1000) are shown in Fig. 3.17.

As can be seen, for all cases of σ_{s2}, the speed of the SAW decreases to the relative limit ($-K^2/2$) for the increasing conductivity of the first layer. However, there are differences depending on the value of the parameter $x = \sigma_{s2}/\sigma_{s1}$. If $x < 1$, which

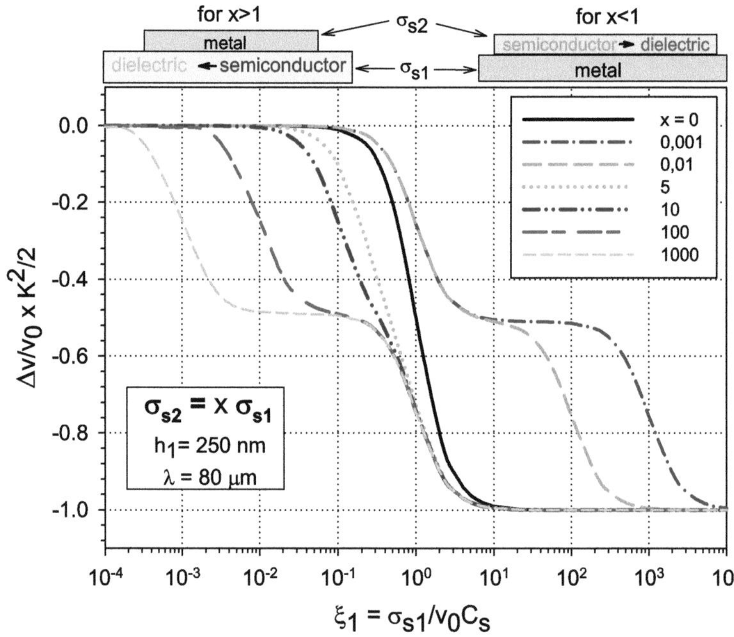

Fig. 3.17 SAW velocity change versus normalized surface conductivity of the first film for seven different values of the surface conductivity of the second film—other remaining parameters: $k_0 = 7.85 \times 10^4\ \text{m}^{-1}$ (for $\lambda = 80\ \mu\text{m}$), $h_1 = 250$ nm

describes a structure in which the second layer has a conductivity less than the first one, then the speed change curve is shifted towards the larger values $\xi_1 > 1$ ($x = 0.001$ is the red curve, $x = 0.01$ is the green ones). On the other hand, for $x > 1$, when the conductivity of the second layer is greater than the first, the trend is the opposite—the speed change curves are shifted towards the smaller values $\xi_1 < 1$. This corresponds to the configuration of semiconductor–metal sensor elements, because as a rule, the conductivity of the metal layers is greater than that of the semiconductor layers. From the point of view of sensor applications, it is advantageous that for this given value of $\xi_1 < 1$, greater velocity changes can be obtained for a structure with a larger parameter x.

However, for larger x-values, e.g. 100 and 1000, describing dielectric-metal structures, there is a range of ξ_1 in which a wide plateau is revealed. In this range, because Δv is constant, independent of the change in conductivity that can be introduced by gas molecules, the working point is shifted along the plateau without changing the wave velocity. The use of acoustoelectric interactions in these cases is very limited, or even impossible. It is easy to see that similar ranges of ξ_1 exist for structures with $x < 1$. These ranges are symmetrical with respect to $\xi_1 = 1$. The structure with $x = 0$ (black curve) represents the ideal dielectric layer on the conductive layer, for which the normalization of velocity changes should be carried out without the geometric factor $1 + \exp(-2k_0h_1)$, as it plays no role in the case of a single layer.

The results of the numerical analysis for the relative changes in attenuation (formula 3.69) are shown in Fig. 3.18. The maximum acoustoelectric attenuation for all x-values is smaller and less sharp compared to the single layer for which it is $K^2/4$ ($x = 0$—black curve). As for velocity, the normalization in this case was performed for $h_1 \to \infty$. The larger the value of the parameter x, the smaller is the maximum attenuation value, aiming for ~0.25. For very small ($x = 0.001$ and 0.01) and very large ($x = 100$, 1000) values of the parameter x, there are oscillations of acoustoelectric damping—the two maximum attenuations are visible. The minimum relative attenuation values which are favorable for excitation and maintenance of oscillation in the sensor system at $0.1{\cdot}K^2/2$ for $\xi_1 =$ ~0.1 and $x = 100$ (pink curve M point) correspond to the plateau area in Fig. 3.17 where there is no significant change in SAW velocity. For bilayer structures with such parameters, there are no large acoustoelectric interactions. On the other hand, a promising solution seems to be structures for which $x =$ ~10 (blue curve), when the conductivity of the second layer is about ten times greater than the first. A fairly large range of values of ξ_1 can then be obtained, for which the acoustoelectric attenuation is almost constant, allowing stable oscillations, even in the case of changes in the conductivity of a first layer due to interaction with a gas.

The bilayer structures, compared to single layers, allow new ranges of normalized conductivity of the first layer ξ_1 to be obtained, depending on the x-parameter, for which there are significant changes in the speed of the SAW. In addition, much lower attenuation values and reduced sharpness of their maxima are characteristic. These are extremely advantageous features from the point of view of sensor applications, as the use of acoustoelectric interactions for these types of structures is more efficient.

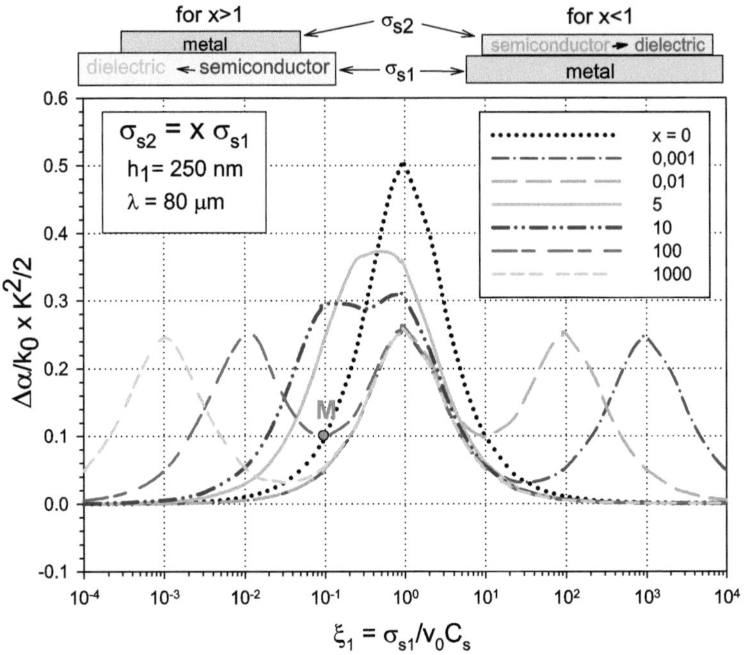

Fig. 3.18 SAW attenuation change versus normalized surface conductivity of the first film for seven different values of the second film surface conductivity—other parameters $k_0 = 7.85 \times 10^4\ \text{m}^{-1}$ (at $\lambda = 80\ \mu\text{m}$), $h_1 = 250$ nm

Similar curves as shown in Figs. 3.17 and 3.18 can be obtained for other commonly used h_1 values, e.g. $h_1 = 50$ nm and $h_1 = 750$ nm. The choice of these values was dictated by the thickness range of the resulting sensor layers. Some subtle differences in the relative changes in velocity and attenuation for the $x = 10$ configuration are shown in Fig. 3.19.

The influence of the thickness of the first layer h_1 in the analyzed range of 50 to 750 nm on the course of the speed changes is extremely small. The differences are barely noticeable in the range $\xi_1 = 0.1–1$, with slightly larger changes occurring for the structure with a smaller thickness h_1 (red curve for $h_1 = 50$ nm). In the case of attenuation, on the other hand, the trend is the opposite—it is smaller for the structure with the largest thickness h_1 (green curve for $h_1 = 750$ nm). This is understandable because the second layer is then further away from the surface of the piezoelectric crystal.

Similar, but inverse effects occur for structures with smaller x-values, e.g. $x = 0.1$, which is shown in Fig. 3.20. The velocity change curves almost overlap, lying to the right of the curve for a single structure (black color), with slightly larger changes visible for the largest thickness of 750 nm. On the other hand, the attenuation, conversely, is smaller for a structure with the smallest thickness of 50 nm. Structures for which $x = 0.1$ are characterized by ten times less conductivity of the second layer

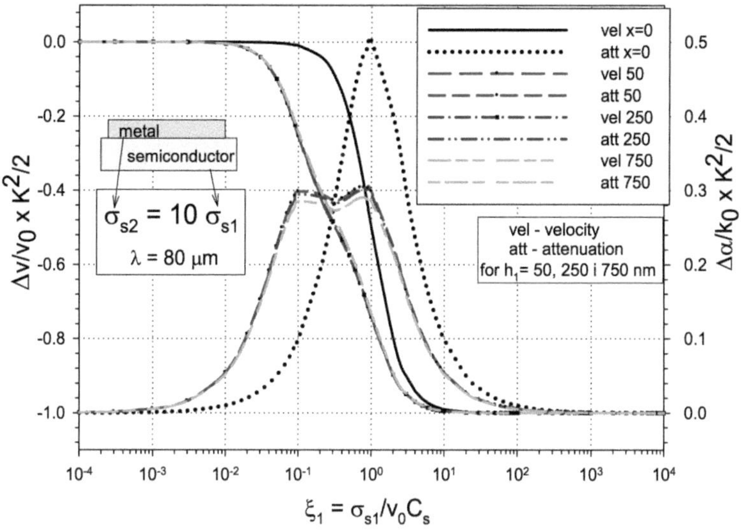

Fig. 3.19 The influence of the three different thicknesses of the first film $h_l = 50$, 250, and 750 nm on a velocity and attenuation changes of bilayer structure with a parameter $x = 10$—the curves in black are for the single film structure with $x = 0$

relative to the first one, so this is an inverse configuration, e.g. metal–semiconductor type.

In the case of a structure for which $x = 1000$ (Fig. 3.21), representing a dielectric-metal system, the effect of the h_1 parameter on the course of the velocity changes is more visible, and it occurs particularly in the $\xi_1 = 0.0035$–0.3 range, where there are small changes in the SAW velocity. As for $x = 10$, larger changes in velocity are visible for the structure with the smallest thickness of 50 nm (color red). On the other hand, the maximum values of the acoustoelectric attenuation, both of the first and second maximums, are slightly smaller for the structures with the largest thickness of 750 nm.

The dispersion effects, i.e. the dependence of both the attenuation and the speed of the SAW on the wavelength, are seen in Expressions (3.69) and (3.70), by the factors $\exp(-2k_0h_1)$ or $1 + \exp(-2k_0h_1)$, where $k_0 = 2\pi/\lambda$. Important is the product of the wave number k_0 and the thickness of the first layer h_1. For the constant value $h_1 = 250$ nm; the typical parameter $x = 10$; and the three SAW wavelengths 8, 80, and 800 μm, the dispersion effects are shown in Fig. 3.22. These effects are similar to the effect of the thickness h_1 on SAW propagation. It is important the reducing of the attenuation for the smallest wavelength $\lambda = 8$ μm. At the same time, the effect of the wavelength on the velocity is smaller.

Analysis in function of normalized conductivity ξ_2 of the second layer

The results of the numerical analysis of the relative velocity changes for the most commonly used values $k_0 = 7.85 \times 10^4\ \text{m}^{-1}$ (at $\lambda = 80$ μm) and $h_1 = 250$ nm, as a

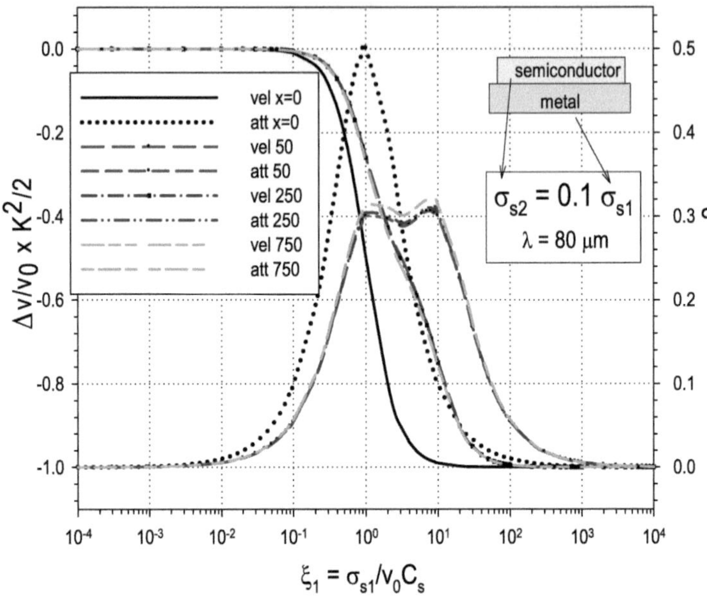

Fig. 3.20 The influence of the three different thicknesses of the first film $h_1 = 50$, 250 and 750 nm on a velocity and attenuation changes of bilayer structure with a parameter $x = 0.1$—the curves in black are for the single film structure

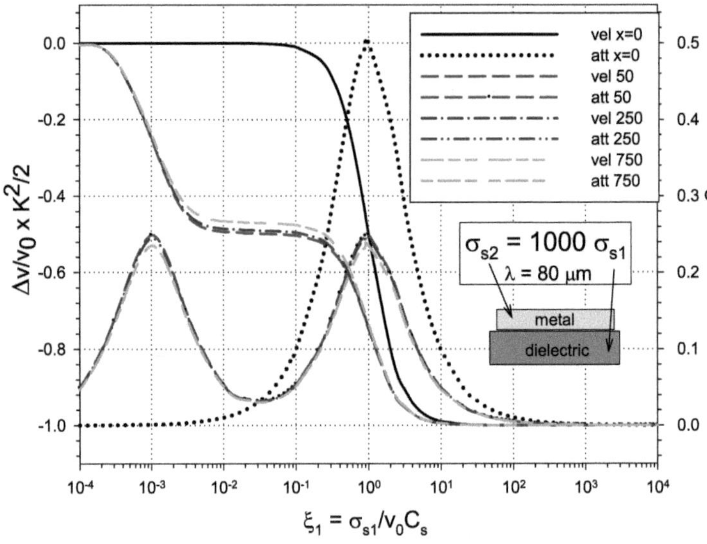

Fig. 3.21 The influence of the three thicknesses of the first film $h_1 = 50$, 250, and 750 nm on a velocity and attenuation changes of bilayer structure with a parameter $x = 1000$—the curves in black are for the single film

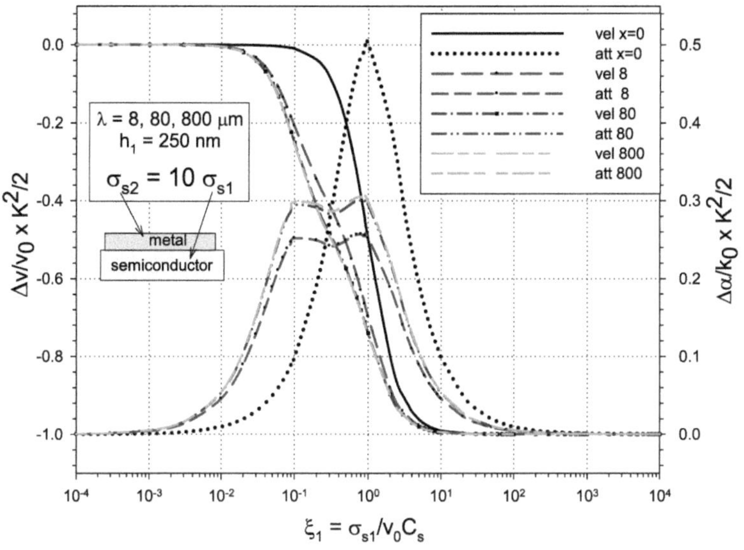

Fig. 3.22 Dispersion effects for three SAW wavelengths 8, 80, and 800 μm and $h_1 = 250$ nm with $x = 10$

function of the normalized surface conductivity of the second layer (parameter $\xi_2 = \sigma_{s2}/v_o C_s$) and seven conductivity values σ_{s1} relative to σ_{s2} ($\sigma_{s1} = x\,\sigma_{s2}$, where $x = 0$, 0.001, 0.01, 5, 10, 100, and 1000) are presented in Fig. 3.23.

As for the relation of $\Delta v/v_0$ in the function of the parameter ξ_1, for all conductivity values of the first layer σ_{s1} relative to the second σ_{s2}, the speed of the SAW decreases to $(-K^2/2)$. If $(\sigma_{s1}/\sigma_{s2}) = x < 1$, then the velocity curves are shifted towards larger values $\xi_2 > 1$ ($x = 0.001$ red curve, $x = 0.01$ green). On the other hand, for $x > 1$, the trend is the opposite—the velocity curves are moved towards smaller values $\xi_2 < 1$. Inthe range of thicknesses $h_1 = 50$–750 nm, there are no significant differences in the course of the velocity changes compared to the corresponding dependencies of ξ_1. On the other hand, the differences appear in the case of $x = 0$, which represents the ideal dielectric in the area of the first layer. The greater the thickness of h_1, i.e. the second layer is further from the piezoelectric waveguide, then the reduction in the propagation speed of the SAW is smaller in the range of $\xi_2 > 1$.

The changes in the SAW acoustoelectric attenuation relative to the normalized surface conductivity of the second layer (parameter ξ_2) are shown in Fig. 3.24. These dependencies are analogous to the obtained curves of attenuation in the function of parameter ξ_1 for the first layer (Fig. 3.18). Slight differences can be seen in the analysis of the amplitude of individual maxima. Compared to the attenuation dependence (Fig. 3.18) all the main maxima for $\xi_2 = 1$ undergo a slight decrease, while the lateral maxima for $x = 0.001$, 0.01, 100, and 1000 increase. The situation in which $x = 0$ corresponds depending on the parameter ξ_2 to configuration: ideal dielectric-dielectric, semiconductor, or metal.

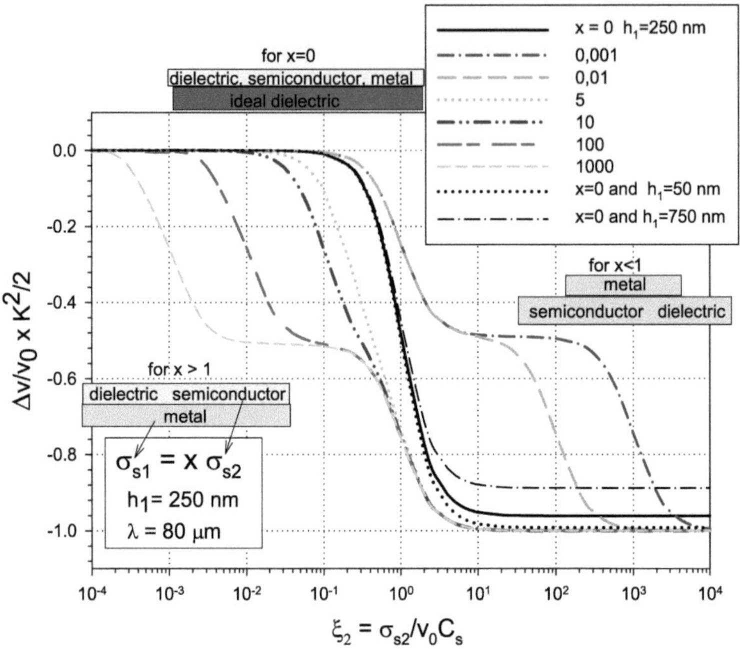

Fig. 3.23 SAW velocity changes versus normalized surface conductivity of the second film for seven different values of the first film surface conductivity: other parameters $k_0 = 7.85 \times 10^4\ \text{m}^{-1}$ (for $\lambda = 80\ \mu\text{m}$), $h_1 = 250$ nm

3.5.4.3 The Use of the Acoustoelectric Effect

The specific layered sensor structure is characterized by particular parameters x and ξ, which defines the working point on the velocity and attenuation curves (see Figs. 3.17 and 3.18). The position of the working point on curves $\Delta v/v_0$ and $\Delta \alpha/k_0$ determines the use of acoustoelectric interactions in gas sensors with SAW. In addition, changes in the electrical conductivity of the structure must also occur as a result of interactions with vapor or gas particles. The working point is the better, the smaller changes in the conductivity of the structure, resulting in larger changes in velocity, which leads to greater acoustoelectric sensitivity. In contrast to the mass sensitivity defined by formula (2.2), the acoustoelectric sensitivity (AE) refers to changes in the electrical conductivity of the sensor structure.

I. *AE sensitivity of single sensor structures* [31]

For single structures, AE sensitivity can be defined as

$$S_{AE} = \lim_{\Delta\sigma_s \to 0} \frac{1}{v_0} \frac{\Delta v}{\Delta \sigma_s} \tag{3.71}$$

By assigning a derivative of Expression (3.43) to σ_s, we get on the basis (3.71):

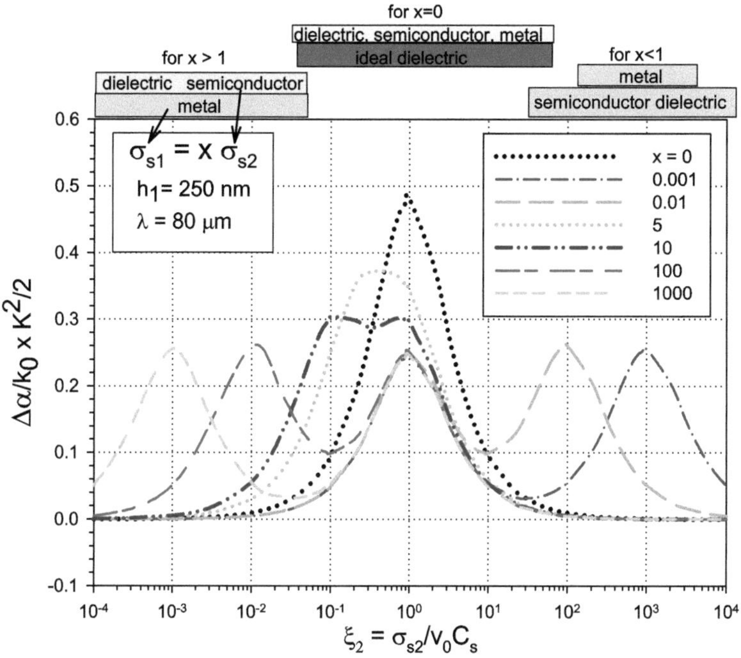

Fig. 3.24 SAW attenuation changes versus normalized surface conductivity of the second film for seven different values of the first film surface conductivity: $k_0 = 7.85 \times 10^4$ m^{-1} (at $\lambda = 80$ μm), $h_1 = 250$ nm

$$S_{AE}(\sigma_s) = -\frac{K^2 v_0^2 C_s^2 \sigma_s}{\left(\sigma_s^2 + v_0^2 C_s^2\right)^2} \tag{3.72}$$

For single structures, only one acoustoelectric parameter $\xi = \sigma_s/v_0 C_s$ is relevant, resulting from the ratio of the electrical surface conductivity of the layer to the parameters of the piezoelectric substrate and the structure environment. The absolute value of acoustoelectric sensitivity can be expressed as follows:

$$|S_{AE}(\xi)| = \frac{K^2}{v_0 C_s} \frac{\xi}{\left(1 + \xi^2\right)^2} \tag{3.73}$$

The graph of this dependency, for the most commonly used piezoelectric substrate $LiNbO_3$ Y–Z, as well as quartz S-T and langazite, using the data from Table 2.1, is presented in Fig. 3.25. The acoustoelectric sensitivity of individual sensor elements increases for small values of the parameter ξ, reaching the maximum for $\xi_{\max} = \sqrt{3}/3 = $ ~0.6, then rapidly decreasing to zero already for ξ ~ 10. This explains the very low acoustoelectric sensitivity of sensor structures characterized by the parameter $\xi > 10$, e.g. thin metal layers, for which depending on their continuity and

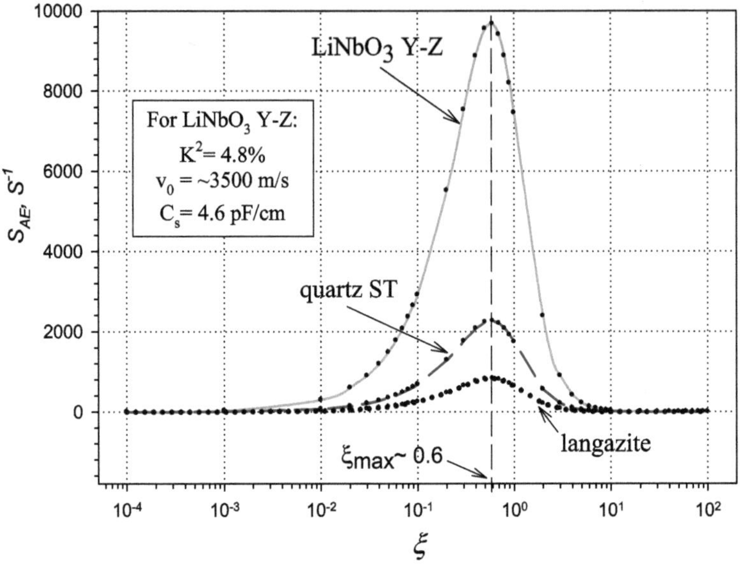

Fig. 3.25 Dependence of the absolute value of the acoustoelectric sensitivity versus parameter $\xi = \sigma_s/v_0 C_s$ of the single sensor structure and three different piezoelectric substrates

thickness $\xi \sim 10^4$–10^5. The location of the maximum acoustoelectric sensitivity does not depend on the parameters of the piezoelectric substrate and environment (K^2, v_0, C_s). On the other hand, its maximum value, which is the largest for $LiNbO_3$ Y–Z substrate, depends on them:

$$S_{AE}^{max} = S_{AE}\left(\xi = \frac{\sqrt{3}}{3}\right) \approx 9682.4\ S^{-1} \tag{3.74}$$

II. *AE Sensitivity for Bilayer Sensor Structures* [31, 42]

For bilayer structures, acoustoelectric sensitivity is a function of the conductivity of both the first and second layers. By determining derivative of Expression (3.70) relative to the conductivity of the first layer σ_{s1} and using (3.71), one can obtain

$$S_{AE}^{(1)} = -\frac{K^2}{(1+e^{-2k_0 h_1})v_0 C_s}\frac{\xi_1\left[1+\xi_2^2(2+\xi_2^2)\right]}{\left[1+\xi_1^2+\xi_2^2+\xi_1^2\xi_2^2\right]^2} \tag{3.75}$$

where $\xi_1 = \frac{\sigma_{s1}}{v_0 C_s}$, $\xi_2 = \frac{\sigma_{s2}}{v_0 C_s}$

The absolute value of $S_{AE}^{(1)}$ determined by dependence (3.75) for $LiNbO_3$ Y–Z substrate, for determined values $h_1 = 250$ nm and $k_0 = 7.85 \times 10^4$ m^{-1} (corresponding to wavelength $\lambda = 80$ μm), as a function of the normalized conductivity of the first layer ξ_1 and the two extremely different conductivity of the second layer

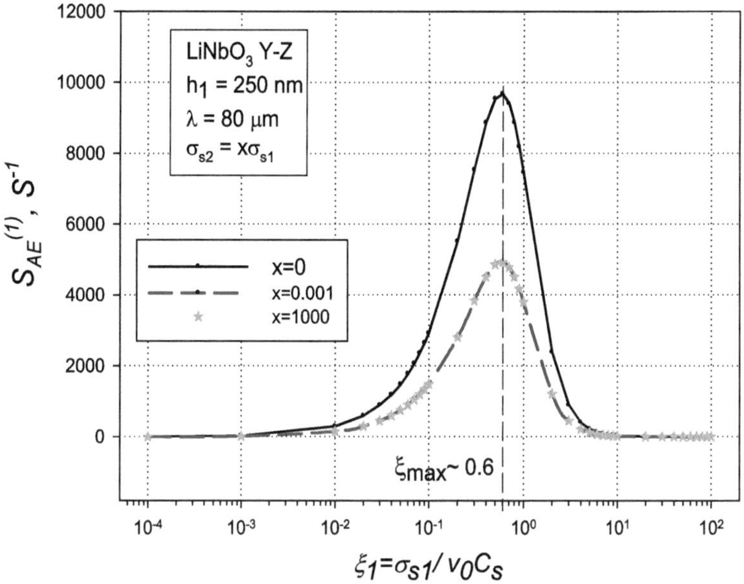

Fig. 3.26 Acoustoelectric sensitivity of the bilayer structures as a function of the ξ_1 parameter of the first layer, for two extreme different values σ_{s2} with respect to σ_{s1}

$x = 0.001$ and 1000, is shown in Fig. 3.26. The acoustoelectric sensitivity of $S_{AE}^{(1)}$ is independent of $x = \sigma_{s2}/\sigma_{s1}$ in a wide range of its values from 0.001 to 1000, except in the case of $x = 0$. It follows that any changes in conductivity of the first layer as a result of interaction with gases will not affect the acoustoelectric sensitivity. The $x = 0$ case is a configuration in which the second layer is the ideal dielectric ($\sigma_{s2} = 0$) and at the same time it is displaced infinitely far from the surface of the piezoelectric substrate. This corresponds to the situation of a single sensor structure.

The maximum AE sensitivity, on the other hand, is slightly dependent on the thickness of the first layer, as shown in Fig. 3.27.

Graphs of the $S_{AE}^{(1)}$ sensitivity defined by formula (3.75), for the determined values $h_1 = 250$ nm and $k_0 = 7.85 \times 10^4$ m^{-1} ($\lambda = 80$ μm), but as a function of the normalized conductivity of the second layer and the five selected conductivity of the first layer, are shown in Fig. 3.28. The conductivity ranges of the first layer correspond to the most typical configurations of bilayer structures.

The $S_{AE}^{(1)}$ sensitivity is also a function of the conductivity of the second layer. Depending on the parameter x, which determines the relationship between the two layers $\sigma_{s1} = x\,\sigma_{s2}$, similar curves of the AE sensitivity curve are obtained, but they are shifted relative to each other. The smaller the parameter x, the more the sensitivity curve is shifted towards the larger values ξ_2, i.e. the greater conductivity of the second layer. The values $x < 1$ represent a sensor structure in which the first layer has less electrical conductivity than the second. This corresponds to configurations of semiconductor–metal structures or dielectric-metal structures. The case $\sigma_{s1} = 0$

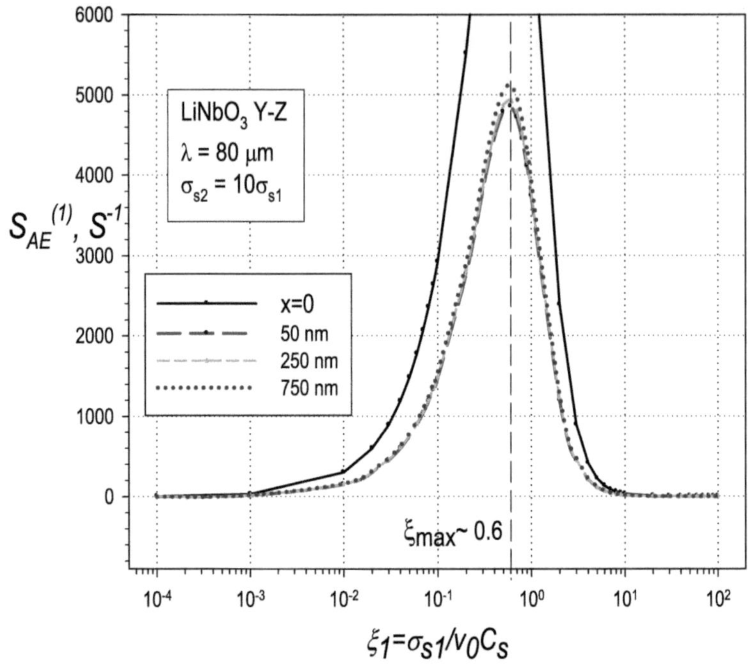

Fig. 3.27 Acoustoelectric sensitivity of the bilayer structure as a function of the ξ_1 parameter of the first layer for its three different thicknesses

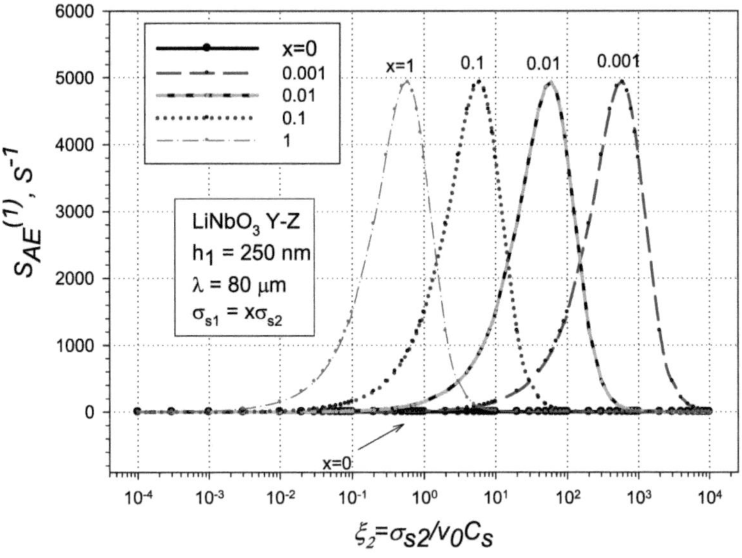

Fig. 3.28 Acoustoelectric sensitivity of the bilayer structure as a function of the ξ_2 parameter of the second film, for five different values $x = \sigma_{s1}/\sigma_{s2}$

describes the effect of the ideal dielectric layer in the area of the first layer, the AE sensitivity is then zero. For $x = 1$, that is, in a situation of equal layer conductivity, the maximum AE sensitivity is for the same value $\xi_2 = \sim 0.6$ as for the single structure.

Using the maximum AE sensitivity of bilayer structures requires matching the *working point* of the structure to its design parameter x, which is defined in the fabrication processes. Based on the case in Fig. 3.28, the position of maximum AE sensitivity is read as a function of the parameter $x = \sigma_{s1}/\sigma_{s2}$, which is shown in Fig. 3.29. For smaller x-values, i.e. smaller surface conductivity of the first layer relative to the second layer, which corresponds to the used bilayer configurations of the type dielectric-metal or semiconductor–metal, to achieve maximum AE sensitivity requires larger values ξ_2, i.e. greater conductivity of the second layer relative to the $v_0 C_s$.

As already mentioned, in the case of bilayer structures, AE sensitivity is also a function of the surface conductivity of the second layer. By assigning derivative of Expression (3.70) relative to σ_{s2}, we obtain an expression similar to (3.75) with an additional factor $\exp(-2k_0 h_1)$:

$$S_{AE}^{(2)} = -\frac{K^2}{(1 + e^{-2k_0 h_1}) v_0 C_s} \frac{\xi_2 \left[1 + \xi_1^2 (2 + \xi_1^2)\right]}{\left[1 + \xi_1^2 + \xi_2^2 + \xi_1^2 \xi_2^2\right]^2} e^{-2k_0 h_1} \tag{3.76}$$

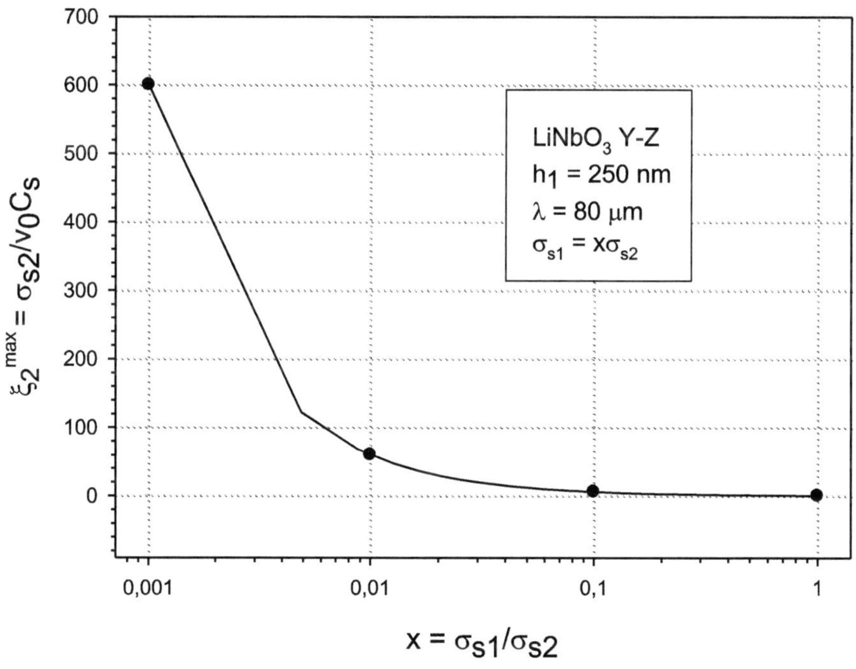

Fig. 3.29 Maximum position of the $S_{AE}^{(1)}$ sensitivity as a function of the structural parameter $x = \sigma_{s1}/\sigma_{s2}$ of the bilayer structure

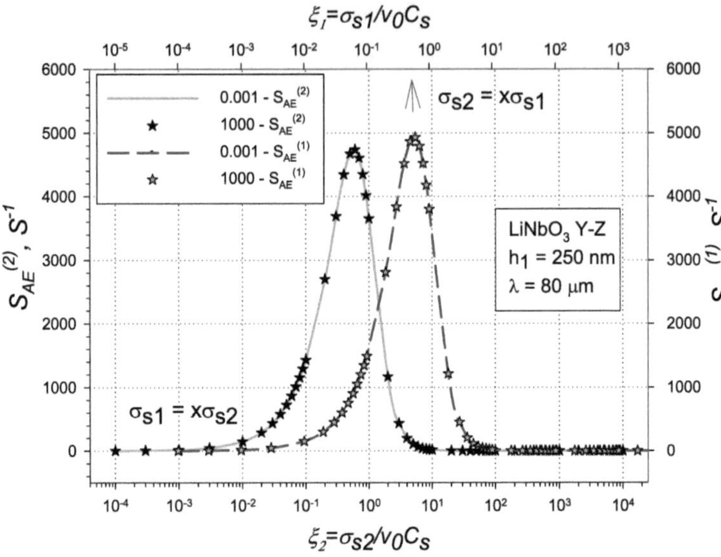

Fig. 3.30 Acoustoelectric sensitivity comparison of the bilayer structure as a function of ξ_1 and ξ_2 parameters, for two extreme different values $x = 0.001$ and 1000

The AE sensitivity with respect to the electrical parameter of the second layer $S_{AE}^{(2)}$ is therefore slightly lower compared to the sensitivity of $S_{AE}^{(1)}$ precisely because of this factor, as can be seen in Fig. 3.30. For better clarity, an additional axis has been introduced for $S_{AE}^{(1)}$ shifted relative to the axis ξ_2. The maximum acoustoelectric sensitivity of $S_{AE}^{(2)}$ with respect to ξ_2 is approximately 200 S^{-1} smaller compared to $S_{AE}^{(1)}$.. Considering the actual AE sensitivity values (~5000 S^{-1}) that are achieved for substrates with high K^2coefficients, the following approximation can be made (for thin sensor layers and not very high SAW frequencies):

$$S_{AE}^{(1)}(\sigma_{s1}) \approx S_{AE}^{(2)}(\sigma_{s2}) \cong S_{AE}(\sigma_{rs}) \tag{3.77}$$

where σ_{rs} is the resultant electrical conductivity of manufactured sensor structures for single or layered samples, obtained using direct measurements by planar electrical method.

3.6 Summing-Up

This chapter presents theoretical models of operation of gas sensors with SAW, for two basic couplings: gas molecules-sensor element and sensor element-SAW.

The first coupling is discussed on examples of thin-film sensor elements most commonly used in SAW sensors, i.e. metal oxides, phthalocyanines, palladium layers, as well as prototype bilayer structures. The result of this basic coupling is changes in the mechanical and/or electrical parameters of these elements. The values of these changes, in turn, determine the mass and/or acoustoelectric sensitivities that can be obtained.

A separate issue was to consider the influence of various types of thin-film sensor structures on the surface of the piezoelectric waveguide on the SAW propagation. For this purpose, perturbation theory was used. According to the principle of small perturbation, the total effect of perturbation of the SAW constant propagation can be considered as the sum of the component perturbations from the individual groups of parameters of the sensor structures. Therefore, coupling models, such as parameters of sensor elements-SAW, separately for their mechanical, viscoelastic, and electrical parameters were presented. Individual sensor structures were considered, as well as for the first time various types of bilayer structures, e.g. semiconductor–metal, for which finite thicknesses and electrical conductivity of each layer were assumed.

For a single structure, an approximate expression is obtained consisting of two components—a negative one representing the *pure* mass load and a positive one taking into account the elastic parameters of the layer. Both of these components are dependent on the wave frequency.

Using the basic equations for electric current density and continuity, and electrostatic approximation, the impedance of a thin semiconductor layer can be determined, taking into account all its electrical parameters (σ, D_e, and ε). For thin semiconductor layers with amorphous structures (e.g. some phthalocyanines), the surface conductivity of the layer is decisive for the disturbance of SAW propagation. The influence of the other electrical parameters (constant diffusion D_e and ε) is negligible.

In the case of bilayer structures, the expressions obtained by the author on the relative changes in SAW attenuation and velocity are the original result of the modification of the energy approach of the subject. The author assumed that the two layers are mutually interconnected through the induced electrical charges in each of them as a result of the electrical potential associated with the SAW. The obtained additional components, compared to the patterns for the single structure, represent the influence of the second layer on the SAW propagation. Depending on the relationship of electrical conductivity of the layers (parameter x), the author conducted a detailed analysis (using the program *Sigma Plot 11.0*) of the changes in propagation of SAW velocity and attenuation, both as a function of the normalized conductivity of the first and second layers.

The analysis showed that, depending on the mutual relation of the layers' electrical surface conductivities, it is possible to achieve significant changes in SAW velocity, as well as smaller attenuation values, compared to single structures, for completely new ranges of normalized conductivity of the first and second layers. Such features are extremely beneficial in terms of obtaining greater acoustoelectric sensitivities and excitation and maintenance of oscillations in the electronic oscillator system of the SAW sensor.

On the basis of the analysis of speed and attenuation changes of SAW for bilayer sensor structures, several of their special properties can be distinguished in comparison with single layers:

- New ranges of normalized conductivity of the first layer ξ_1, depending on the ratio of surface electrical conductivity of both layers (parameter $x = \sigma_{s2}/\sigma_{s1}$), in which significant changes in SAW speed can be obtained.
- For larger values of the parameter x, these ranges shift towards smaller values ξ_1, and vice versa.
- The values of the parameter ξ_1, for which there are maximum speed changes, are inversely proportional to x; $\xi_{1\max} = 1/x$, e.g. for $x = 10$, $\xi_{1\max} = 0.1$ and 1, while for $x = 0.001$, $\xi_{1\max} = 1$ and 10^3 are obtained.
- The width of the plateau area in which there is no significant change in velocity for $K^2/4$ increases both when the parameter x increases and when it decreases.
- The maximum acoustic attenuation is smaller and less sharp for all values of the parameter x.
- The oscillation character of attenuation changes is characteristic for both very small and very large values of the parameter x.
- Minimum attenuation values correspond to plateau areas where there is no change in SAW velocity.
- The influence of the thickness of the first layer h_I, in the analyzed range 50–750 nm, on the course of changes in speed and SAW attenuation is extremely small—slightly larger changes were found for structures with a very large parameter $x = 1000$, which represents the structure-type dielectric-metal.
- The dispersion effects, i.e. the dependence of the speed and attenuation of SAW wavelength, are similar to the influence of the thickness of the first layer h_I—it is important to reduce the attenuation for smaller wavelengths.

The author performed an analysis of acoustoelectric sensitivity for single and double sensor structures as a function of their normalized electrical conductivity ξ.

For single structures, it can be stated that

- The absolute AE sensitivity value increases for small values of the parameter ξ, reaching the maximum for $\xi_{\max} = \sqrt{3}/3 = $ ~0.6, then rapidly decreasing to zero for ξ ~ 10.
- The position of this maximum is independent of the piezoelectric material used, while the maximum value depends on the electromechanical coupling coefficient of the substrate K^2.

In the case of two-layer structures, AE sensitivity is a function of the electrical conductivity of the first and second layers. This sensitivity is smaller compared to single structures, but its maximum position depends on the x-parameter ($\sigma_{s1} = x\,\sigma_{s2}$) for different values of the normalized conductivity of the second layer.

Using the maximum AE sensitivity of bilayer structures requires matching the *working point* of the structure to its design parameter x, which is defined in the fabrication processes. For smaller x-values, i.e. lower surface conductivity of the first layer relative to the second, which corresponds to the used bilayer configurations of the type dielectric-metal or semiconductor–metal, to achieve maximum AE sensitivity requires larger values of ξ_2 parameter, i.e. greater conductivity of the second layer relative to the substrate parameters (i.e. $v_0 C_s$).

References

1. Adamson AW (1982) Physical chemistry of surfaces, 4th edn. Wiley, New York
2. Anisimkin VI (1996) Russia-Italy researches on surface acoustic wave (SAW) gas sensors. In: IEEE ultrasonic symposium, pp 671–673
3. Arbiol J, Morante JR, Bouvier P, Pagnier T, Makeeva EA, Rumyantseva MN, Gaskov AM (2006) SnO_2/MoO_3-nanostructure and alcohol detection. Sens Actuators B 118:156–162
4. Auld BA (1973) Acoustic fields and waves in solids. II, ch 12, Wiley, New York
5. Ballantine DS, White RM, Martin SJ, Ricco AJ, Frye GC, Zellers ET, Wohltjen H (1997) Acoustic wave sensors: theory, design and physico-chemical applications. Academic Press, New York
6. Barsan N, Weimar U (2001) Conduction model of metal oxide gas sensors. J Electroceram 7:143–167
7. Bers A (1970) Interaction between acoustic surface waves and electrons in solids. In: Proceedings of ultrasonic symposium, San Francisco, California, pp 138–172
8. Bielański A (1963) Chemia fizyczna (Physical chemistry). PWN, ch 14, p 487
9. Bott B, Jones TA (1984) A highly sensitive NO_2 sensor based on electrical conductivity changes in phthalocyanine films. Sens Actuators 5:43–53
10. Brattain W, Bardeen J (1953) Surface properties of germanium. Bell Syst Tech J 32
11. Brunnauer S (1945) The adsorption of gases and vapours. Oxford University Press and Princeton University Press, Princeton NJ
12. Comini E, Sberveglieri G (2010) Metal oxide nanowires as chemical sensors. Mater Today 13(7–8):36–44
13. Comini E, Sberveglieri G, Barreca D, Sada C, Gasparotto A, Maccato Ch, Tondello E (2009) Chemical vapor deposition of Cu_2O and CuO nanosystems for innovative gas sensors. In: IEEE sensors conference, pp 111–113
14. Datta S (1985) Surface acoustic wave devices. Prentice-Hall, Enclewood Cliffs, NJ, ch 7, pp 179–186
15. D'Amico A, Palma A, Verona E (1982) Hydrogen sensor using a palladium coated surface acoustic wale delay-line. In: IEEE ultrasonic symposium, vol 1. San Diego, pp 308–311
16. Fabre A, Finot E, Demoment J, Contreras S (2003) In situ measurement of elastic properties of PdH_x, PdD_x and PdT_x. J Alloy Compd 356–357:372–376
17. Favier F, Walter E, Zach MP, Benter T, Penner R (2001) Hydrogen sensors and switches from electrodeposited palladium mesowire arrays. Science 293:2227–2231
18. Fernandez MJ, Fontecha JL, Sayago I, Aleixandre M, Lozano J, Gutierrez J, Gracia I, Cane C, Horrillo M (2007) Discrimination of volatile compounds through an electronic nose based on ZnO SAW sensors. Sens Actuators B 127(1):277–283
19. Fisher BH, Malocha DC (2010) Study of the acoustoelectric effect for SAW sensors. IEEE UFFC 57(3):698–706
20. Hechner J (1995) Formowanie warstwy ftalocyjaninu miedzi do sensorów czułych na NO_2 (Forming of copper phthalocyanine layer for NO_2 sensitive sensors). Materiały Elektroniczne ITME 23(3):24–41
21. Ingebrigtsen KA (1970) Linear and nonlinear attenuation of acoustic surface waves in a piezoelectric coated with a semiconducting film. J Appl Phys 41:454–459
22. Ippolito SJ, Kandasamy S, Kalantar-Zadeh K, Włodarski W (2005) Layered SAW hydrogen sensor with modified tungsten trioxide selective layer. Sens Actuators B 108:553–557
23. Jakubik WP (2005) Hydrogen detection by single and bilayer sensor structures in surface acoustic wave system. J Phys IV 129:117–120
24. Jakubik WP (2006) Investigations of bilayer sensor structure with copper phthalocyanine and palladium for hydrogen detection in SAW system. J Phys IV 137:95–98
25. Jakubik WP (2007) Investigations of thin film structures of WO_3 and WO_3 with Pd for hydrogen detection in a surface acoustic wave sensor system. Thin Solid Films 515(23):8345–8350
26. Jakubik W (2008) Interactions of the Polyaniline and Nafion bilayer sensor structure with ammonia in a dry and wet air atmosphere. Eur Phys J Spec Top 154(1):93–96

27. Jakubik W (2009) Hydrogen gas-sensing with bilayer structures of WO_3 and Pd in SAW and electric systems. Thin Solid Films 517(22):6188–6191
28. Jakubik W (2011) Surface acoustic wave-based gas sensors. Thin Solid Films 520(3):986–993
29. Jakubik W (2014) Elemental theory of a SAW gas sensor based on electrical conductivity changes in bi-layer nanostructures. Sens Actuators B 203:511–516
30. Jakubik W, Krzywiecki M, Maciak E, Urbańczyk M (2012) Bi-layer nanostructures of CuPc and Pd for resistance-type and SAW-type hydrogen gas sensors. Sens Actuators B. https://doi.org/10.1016/j.snb.2012.10.056
31. Jakubik W, Powroźnik P, Wrotniak J, Krzywiecki M (2016) Theoretical analysis of acousto-electrical sensitivity in SAW gas sensors with single and bi-layer structures. Sens Actuators B 236:1069–1074
32. Jakubik W, Urbańczyk M (1997) The electrical and mass effect in gas sensors of the SAW type. J Tech Phys 38(3):187–195
33. Jakubik WP, Urbańczyk MW (2005) SAW hydrogen sensor with a bilayer structure based on interaction speed. Sens Actuators B 106(2):602–608
34. Jakubik WP, Urbańczyk MW, Bethanabotla V, Cular S (2006) Acoustoelectric effect in surface acoustic wave gas sensors (SAW) with metal-free phthalocyanine and palladium sensing bi-layers. ECS Trans 1(19):1–9
35. Jakubik WP, Urbańczyk MW, Bodzenta J, Pietrzyk MA (2005) Investigations on the resistance of the metal-free phthalocyanine and palladium bilayer sensor structure influenced by hydrogen. Sens Actuator B 105(2):340–345
36. Jakubik WP, Urbańczyk MW, Kochowski S, Bodzenta J (2002) Bilayer structure for hydrogen detection in a surface acoustic wave sensor system. Sens Actuators B 82(2–3):265–271
37. Jakubik WP, Urbańczyk MW, Kochowski S, Bodzenta J (2003) Palladium and phthalocyanine bilayer films for hydrogen detection in a surface acoustic wave sensor system. Sens Actuators B 96(1–2):321–328
38. Jakubik WP, Urbańczyk MW, Maciak E (2006) Palladium and metal-free phthalocyanine bilayer structures for hydrogen detection in the SAW sensor system based on interaction speed. IEEE Sens J 6(5):1178–1185
39. Jakubik W, Urbańczyk M, Maciak E, Krzywiecki M (2011) Bi-layer nano-structures of CuPc and Pd in SAW and resistance hydrogen sensor. Procedia Eng 25:252–255
40. Jakubik WP, Urbańczyk M, Maciak E, Pustelny T (2009) Bilayer structures of NiO_x and Pd in surface acoustic wave and electrical gas sensor systems. Acta Phys Pol A 116(3):315–320
41. Jakubik W, Urbańczyk M, Opilski A (2001) Sensor properties of lead phthalocyanine in a surface acoustic wave system. Ultrasonics 39(3):227–232
42. Jakubik W, Wrotniak J, Powroźnik P (2018) Theoretical analysis of a surface acoustic wave gas sensor mechanism using electrical conductive bi-layer nanostructures. Sens Actuators B 262:947–952
43. Jakubik W (2012) Theory of SAW gas sensor based on bi-layer conductivity changes. Procedia Engineering 1287–1290, Proc. Eurosensors XXVI, Kraków, Poland
44. Jakubik W (2012) Zastosowanie akustycznych fal powierzchniowych w czujnikach gazowych (Application of surface acoustic waves in gas sensors). Wydawnictwo Politechniki Śląskiej, Silesian University of Technology Monograph no. 413
45. Jakubik W (1997) Propagacja akustycznej fali powierzchniowej w cienkich warstwach wybranych ftalocyjanin (Propagation of surface acoustic wave in thin films of selected phthalocyanines). Dissertation, Silesian University of Technology, Gliwice
46. Jakubik W, Urbańczyk M, Maciak E (2008) Detekcja wodoru z wykorzystaniem palladu oraz tlenku niklu w układzie sensorowym z akustyczną falą powierzchniową (Hydrogen detection with the use of palladium and Nikel oxide in sensor system with surface acoustic wave). Problemy Eksploatacji 4
47. Janata J, Huber R (1985) Solid state chemical sensors. Academic Press, Orlando
48. Kanari M, Karino Y, Wakamatsu T (2005) Nanoidentation test for evaluating intermolecular elastic and plastic characteristics of CuPc thin films. Jpn J Appl Phys 44:8249–8255

49. Kay B, Peden Ch, Goodman W (1986) Kinetics of hydrogen absorption by Pd(110). Phys Rev B 34(2):817–822
50. Kearns D, Calvin M (1961) Electrical properties of organic solids. Charge carrier diffusivity in metal-free phthalocyanine. J Chem Phys 34(6):2022–2025
51. Kino G, Reeder T (1971) A normal mode theory for the Rayleigh wave amplifier. IEEE Trans ED 18(10):909–920
52. Lakin KM (1971) Perturbation theory for electromagnetic coupling to elastic surface waves on piezoelectric substrates. J Appl Phys 42(3):899–906
53. Lakin K, Shaw A (1969) Surface wave delay line amplifiers. IEEE Trans MTT 17(11):912–920
54. Lewis FA (1967) The palladium hydrogen systems. Academic Press, New York
55. Licznerski B (2004) Thick-film gas microsensors based on tin dioxide. Bull Pol Acad Sci Tech 52(1):37–42
56. Licznerski B, Nitsch K, Teterycz H, Sobański T, Wiśniewski K (2004) Characterisation of electrical parameters for multilayer SnO_2 gas sensors. Sens Actuators B 103:69–75
57. Lundstroem KI, Shivaraman MS, Svensson CM (1975) A hydrogen-sensitive Pd-gate MOS transistor. J Appl Phys 46(9):3876–3881
58. Lundstroem KI, Svensson Ch (1985) Gas-sensitive metal gate semiconductor devices. In: Janata J, Huber R (eds) Solid state chemical sensors, ch III. AP, New York, p 33
59. Martin SJ, Frye GC, Senturia SD (1994) Dynamics and response of polimer-coated surface acoustic wave devices: effect of viscoelastic properties and film resonance. Anal Chem 66(14):2201–2219
60. Martin SJ, Frye GC, Ricco AJ, Zipperian TE (1987) Measuring thin film properties using SAW devices: diffusivity and surface area. In: IEEE ultrasonic symposium (Proc Denver, Colorado), vol 1, pp 563–568
61. Martin SJ, Frye GC, Spates JJ, Butler MA (1996) Gas sensing with acoustic devices. In: IEEE ultrasonic symposium. San Antonio Texas, pp 423–434
62. Maziarz W, Pisarkiewicz T (2008) Gas sensors in a dynamic operation mode. Meas Sci Technol 19(5):055205
63. Mitra P, Mukhopadhyay AK (2007) ZnO thin film as methane sensor. Bull Pol Acad Sci Tech 55(3):281–285
64. Morrison SR (1994) Chemical sensors. In: Sze (ed) Semiconductor sensors. Wiley, New York
65. Moser FH, Thomas AL (1982) The phthalocyanines. In: Properties, vol I. CRC Press Inc., Boca Raton
66. Nieuwenhuizen M, Nederlof A, Barendsz W (1988) Metallopthalocyanines as chemical interfaces on a surface acoustic wave gas sensor for nitrogen dioxide. Anal Chem 60(3):233–239
67. Penza M, Casano G, Tottorella F (2001) Gas recognition by activated WO_3 thin-film sensors array. Sens Actuators B 81:115–121
68. Ricco AJ, Martin SJ, Zipperian TE (1985) Surface acoustic wave gas sensor based on film conductivity changes. Sens Actuators 8:319–333
69. Sakai Y, Sadaoka Y, Yokouchi H (1974) Electrical properties of evaporated thin films of copper phthalocyanine. B Chem So Jpn 47(8):1886–1888
70. Seiyama T, Kato A, Fulishi K, Nagatani M (1962) A new detector for gaseous components using semiconductive thin films. Anal Chem 34:1502–1503
71. Sivaramakrishnan S, Rajamani R, Smith CS, McGee KA, Mann KR, Yamashita N (2008) Carbon nanotube coated surface acoustic wave sensor for carbon dioxide sensing. Sens Actuators B 132:296–304
72. Śmiałowski M (1961) Wodór w stali (Hydrogen in steel). WNT, Warszawa, p 24
73. Snow A (1986) Simultaneous electrical conductivity and piezoelectric mass measurements on iodine-doped phthalocyanine Langmuir-Blodgett films. Langmuir 2:513–519
74. Sze SM (1969) Physics of semiconductor. Wiley, New York
75. Szuber J, Czempik G, Larciprete R, Adamowicz B (2000) The comparative XPS and PYS studies of SnO_2 thin films prepared by L-CVD technique and exposed to oxygen and hydrogen. Sens Actuators B 70:177–181

76. Szuber J, Grządziel L (2001) Photoemission study of the electronic properties of in situ prepared copper phthalocyanine (CuPc) thin films exposed to oxygen and hydrogen. Thin Solid Films 391:282–287
77. Tabib-Azar M, Sutapun B, Petrick R, Kazemi A (1999) Highly sensitive hydrogen sensors using palladium coated fiber optics with exposed cores and evanescent field interactions. Sens Actuators B 56:158–163
78. Taguchi N (1962) Chemical sensor technology. In: Chiba A (ed) 4, 1991 Kodansha, Tokio
79. Thomas RC, Hughes RC (1997) Sensors for detecting molecular hydrogen based on Pd metal alloys. J Electrochem Soc 144(9):3245–3249
80. Thomson M, Stone DC (1997) Surface-launched acoustic wave sensors. Wiley, New York
81. Tiersten H, Shina BKJ (1978) A perturbation analysis of the attenuation and dispersion of surface waves. J Appl Phys 49(1):87–95
82. Urbańczyk M, Jakubik W, Kochowski S (1994) Investigation of sensor properties of CuPc with the use of SAW. Sens Actuators B 22(2):133–137
83. Vaz AR, Salvadori MC, Cattani M (2003) Young Modulus measurement of nanostructured palladium thin films. Nanotech 3:177–180
84. Walter EC, Favier F, Penner R (2002) Palladium mesowire arrays for fast hydrogen sensors and hydrogen-actuated switches. Anal Chem 74:1546–1553
85. Wang W-C, Lin Y (1999) Acousto-electric attenuation in a layered structure analyzed by the transmission line technique. IEEE Trans UFFC 46(3):549–555
86. Williams DE (1999) Semiconducting oxides as gas-sensitive resistors. Sens Actuators B 57:1–16
87. Windishmann H, Mark P (1979) A model for the operation of a thin film SnO_x conductance modulation carbon monoxide sensor. J Electrochem Soc 126:627–633
88. Wohltjen H (1984) Mechanism of operation and design considerations for surface acoustic wave device vapour sensors. Sens Actuators 5:307–325
89. Wright JD (1989) Gas adsorption on phthalocyanines and its effects on electrical properties. Prog Surf Sci 31:1–60
90. Zakrzewska K, Radecka M (2007) TiO_2-SnO_2 system for gas sensing—photodegradation of organic contaminants. Thin Solid Films 515:8322–8338
91. Zhou J, Tao Ch, Yin Ch (2008) Analysis of different effect on the response of SAW hydrogen gas sensor. In: Proceedings of the 3rd IDEE international conference on nano/micro engineered molecular systems, Sanya, China, pp 1168–1171

Chapter 4
Methodology of SAW Gas Sensor Testing

Abstract In this chapter, the most important features of the dual-delay line measuring system often used in practice are described. This system compensates for the influence of some undesirable external factors on the measurement results, e.g. small changes in temperature or pressure. The system enables precise examination (by measuring frequency changes) of different sensor layers and structures made in the SAW system, placed in different gas atmospheres. In order to confirm the acoustoelectric interactions that result from changes in the electrical conductivity of the sensor layer or structure, the planar electrical method was additionally introduced. For sensor structures produced in the same technological processes, it is possible to correlate acoustic and electrical responses, and thus confirm the occurrence of the acoustoelectric interactions.

The methodology of testing gas sensors with SAW is dependent on the design factors, the measurement method used, and the information we want to obtain on the basis of the research. The design factors are primarily the type of sensor module used, i.e. resonator or delay line, and the number of acoustic channels designed (single-, two-, or multi-channel modules). On the other hand, the measurement method is related to the choice of the measured quantity, the changes of which should depend on the type and composition of the gas atmosphere in the sensor environment. These quantities can be amplitude, phase, and frequency of SAW in the oscillator system.

From the point of view of sensor applications, the most important information about the investigated structure is its sensitivity to a particular type of vapor or gas under the specified measurement conditions, i.e. temperature, humidity, or presence of other gases. The magnitude of the changes in the amplitude, phase, or frequency of the SAW must be recorded. On the other hand, when the purpose of the research is more basic information, to determine what kind of interactions (mass and/or acoustoelectric) we are dealing with, more data must be obtained, e.g. to determine whether the speed of the SAW decreases or increases and whether there are changes in the electrical conductivity of the structure. In these cases, either special experiments are used, e.g. with a shorted (compact) electric field on the piezoelectric surface, which eliminate acoustoelectric interactions [4], or complex methods are used, consisting

W. Jakubik, *Application of Surface Acoustic Waves in Gas Sensors*, Springer Series in Surface Sciences 69, https://doi.org/10.1007/978-3-031-89971-3_4

of simultaneous acoustic and electrical investigations of sensor structures performed under the same technological conditions.

The relative changes in the amplitude, phase and frequency of SAW depend on the sensor elements used, the concentrations of vapors or gases, the interaction temperature, the humidity. These changes are usually small. Appropriately sensitive instruments, e.g. spectral analyzers, vector voltmeters, or appropriate measurement methods, e.g. oscillator systems, must be available in laboratory conditions to monitor changes in SAW velocity through oscillation frequency. An additional difficulty may be the influence of external interfering factors, e.g. fluctuations in temperature, pressure, humidity, electric fields, which may affect the attenuation and propagation rate of SAW. External factors can influence both the parameters of the piezoelectric substrate and the parameters of the sensor elements themselves.

Practical solutions for SAW sensors are very often based on *frequency shift* measurements of the respective oscillation systems, as a result of changes in the speed of SAW propagation. Phase shift measurements are rarely used and amplitude change measurements very rarely [1]. This is due both to the properties of the sensor elements used (for many of them, the wave amplitude does not change much, even for large concentrations of gases) and to the ease and high accuracy in frequency measurement in the case of velocity changes.

As mentioned in Chap. 2, SAW gas sensors are most often made in an oscillator system with a delay line, because it naturally has free space to produce sensor elements in the area between the transducers. Single-channel modules were used in the early development of this type of sensor, including in the first SAW sensor [9]. Currently, however, two-channel sensor modules on a common piezoelectric substrate are most commonly used. They form the main element of the very often used double-delay measuring method.

4.1 Preparation of SAW Sensor Module

Several basic steps can be distinguished in the methodology of testing gas sensors with SAW, carried out under laboratory conditions, using oscillation systems. These are preparation of the sensor module, selection of the measurement method, obtaining stable oscillations of the electronic system of the tested sensor module under specific measuring conditions (temperature, gas flow), generation of gas atmosphere with different composition in the measuring chamber by means of dosing system (concentration of gas or gases, humidity of carrier gas), acquisition of frequency changes (differential, differential and/or resonance) in function of dosed gas concentrations.

At the stage of preparation of the sensor module, the piezoelectric substrate, the number of acoustic channels and the geometric dimensions should be selected. The high coefficient of piezoelectric coupling K^2 of the substrate material ($LiNbO_3$ Y–Z) is one of the parameters determining the possibility of obtaining significant acoustoelectric interactions. However, in the case of sensor structures for which this type of interaction is not important, substrates with a lower coefficient K^2 are sufficient. In addition, the limiting parameter is also the Curie T_c point, above which the materials lose their piezoelectric properties. A new substrate type (e.g. langazite) for which T_c ~1200 °C should be used to study the sensor properties of structures at very high temperatures of ~800 °C. Important parameters of the most commonly used piezoelectric substrates in SAW sensors are presented in Table 2.1 (see Chap. 2). Most of them can be obtained at the Institute of Electronic Materials Technology (ITME) in Warsaw, while langazite can be obtained at FOMOS-MATERIALS in Moscow. The dimensions of the piezoelectric plates used in the author's research, for sensor modules with double-delay line, are usually 30 × 20 × 2 mm. The plates were obtained at ITME and are adapted to the universal 24-pin interchangeable metal bases of Kazel, Koszalin, Poland.

On the properly prepared surface of the piezoelectric crystal (appropriate degree of purity), delay lines are made from interdigital transducers. As already mentioned in Chap. 2, the studies used ordinary transducers with a width of $p = 20$ μm and the gaps between them also $p = 20$ μm, which gives the length of the generated SAW $\lambda = 80$ μm ($\lambda = 4p$). The number of electrode pairs of each transducer was $N = 20$. Such transducer parameters ensured good repeatability within the framework of the available photolithography technology.

The next step is to make the appropriate masking. Precisely shield the transducers made, leaving the appropriate place(s) for the sensor structure or structures. Mechanical masks and disposable masks made of thin aluminum foil are used. The area for the sensor structure is usually a rectangle of about 14 × 8 mm in the case of 30 × 20 mm tiles with two delay lines.

Sensor structures are applied to the unmasked area of the substrate using available thin-film technologies. Vacuum vapor, cathode sputtering, and spin and spray coatings technologies are most commonly used. After the structure is made, the mask is removed and electrical connections are fed to the transducers using thin copper wire, conductive adhesive, or ultrasonic bonding method. Ultrasonic wire assembly is used in large-scale production.

In the final stage, the prepared sensor module is placed on a universal metal base, which has previously been supplied with a thick-film heater. The various stages of preparation of the sensor module from SAW are shown in Fig. 4.1.

Depending on the sensor modules used (single-, two-, or four-channel), the oscillation recording method includes: the resonance frequency of the sensor module with a single-delay line, the differential frequency of the module with a double-delay line and the differential frequencies of the individual measurement channels.

In the course of many years of work on gas sensors with SAW, the author has made studies of dozens of sensor structures with different coatings in the form of single layers. On the other hand, layer structures (two- and multilayer) are an important novelty in this type of sensors. In the following part, the most important features of the used measurement methods and systems will be presented and discussed.

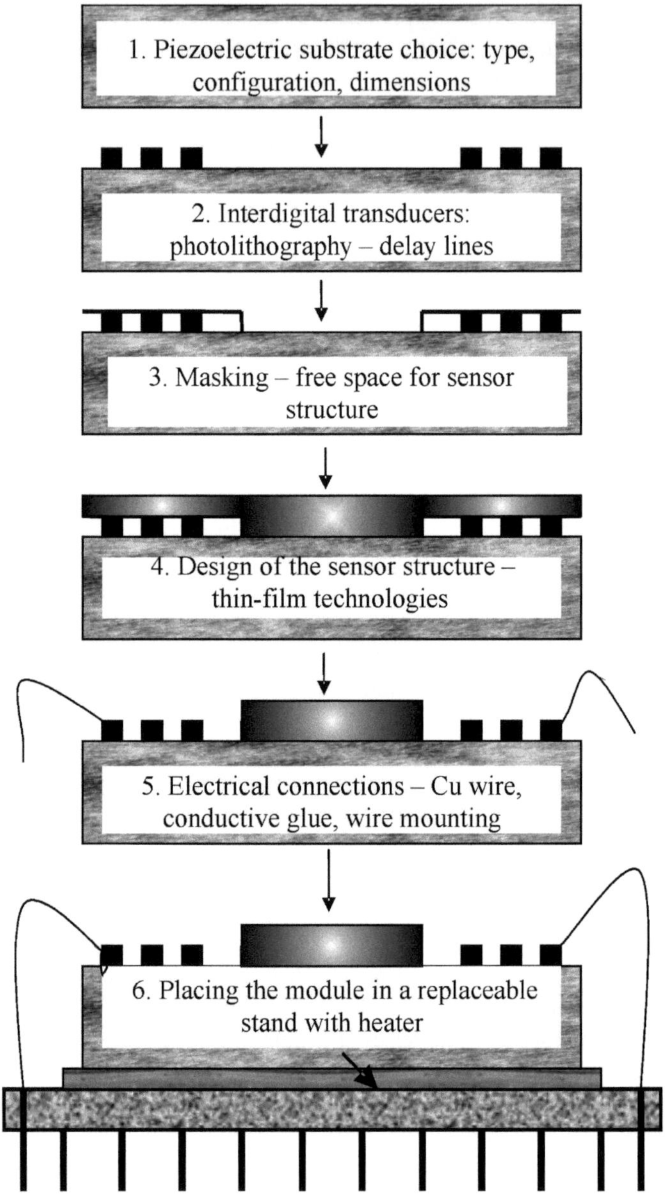

Fig. 4.1 Stages of preparation of sensor module with SAW

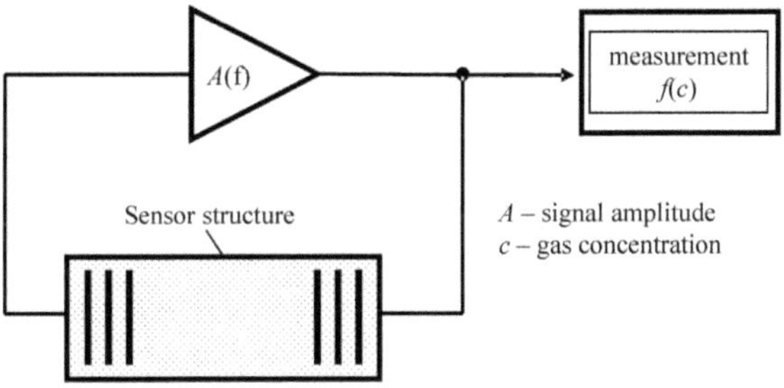

Fig. 4.2 Single-delay line system with a sensor structure

4.2 Measuring Method with Single-Delay Line

A single-delay line as a measurement method was used in the initial phase of development of SAW gas sensors. The substrates can be smaller because only one delay line is needed (Fig. 4.2). The features of the method are: direct measurement of the oscillator frequency (in the MHz range) and no temperature compensation. Obtaining stable oscillations over a long period of time requires very good temperature stabilization. Direct measurement of the resonance frequency gives information about the directions of changes in the speed of the SAW as a result of the interaction of the structure with the gas molecules. The increase in the frequency f means an increase in the speed and vice versa, which results from the oscillator loop principle (see Sect. 2.2.3).

4.3 Measuring Method with Dual-Delay Line

In laboratory studies, the measurement method with a double-delay line, performed on a common piezoelectric substrate, is most commonly used (Fig. 4.3). The method was first proposed by Wohltjen and Dessy [9] and consists of placing each line in the feedback loop of the corresponding amplifier. Using simple electronic circuits, it is possible to obtain independent oscillations f_0 and f of each of the loops and to separate (mixing) their frequency difference Δf.

The method, in application to gas sensors, has two advantageous features due to the use of a reference line (without coverage). It provides compensation for adverse changes in temperature and pressure surrounding the atmosphere and reduces the measurement frequency from the range of MHz (for resonant frequencies) to kHz (for the frequency Δf separated from the mixer). These features are extremely important in practical applications. In addition, the use of a delay line makes it possible to conveniently mask the substrate and produce a sensor structure in one of the tracks, since there is sufficient space between the transmitter and receiver. Both delay lines are placed in independent generation circuits, which enable the excitation of SAW with a frequency determined by the width of the transducers, the properties of the

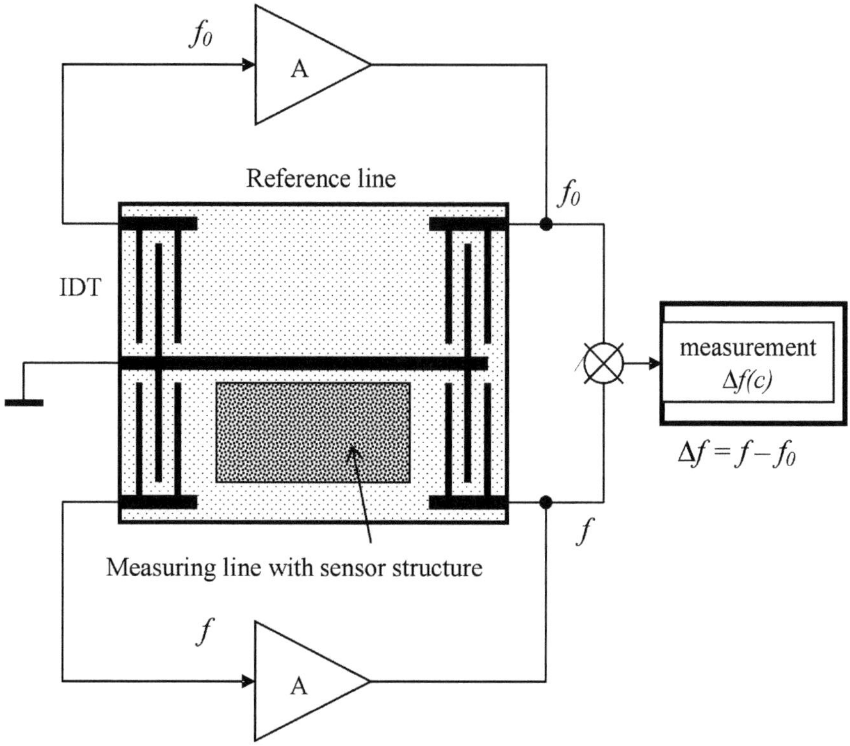

Fig. 4.3 Dual-delay line method, IDT—interdigital transducers for SAW generation, A—amplifier

piezoelectric substrate, and the parameters of the sensor structure (density, elastic parameters, and surface electrical conductivity).

The principle of operation, in electronic terms, of a single-delay line in the oscillator system was presented earlier (see Sect. 2.2.3, Fig. 2.6, Eqs. (2.5)–(2.6)). The simple consequence of the phase condition (2.6) is that for each oscillator the frequency of the oscillation is directly proportional to the velocity of the excited surface wave. In the case of a measurement line with a sensor structure, in which oscillations of frequency f are excited, while the velocity of SAW has a value v, we get [7]:

$$f = \frac{2\pi n_0 - \phi_{ext}}{2\pi L} v \tag{4.1}$$

where $\phi_{ext} = \phi_A + 2\phi_{tr}$ is the phase shift from the amplifier and two transducers, L is the length of the delay line, and n_0 is the ratio of the length of the delayed line to the length of the excited wave.

On the other hand, the difference in frequency, $\Delta f = f - f_0$, both acoustic lines is

$$\Delta f = \frac{2\pi n_0 - \varphi_{ext}}{2\pi L}\Delta v \tag{4.2}$$

where $\Delta v = v - v_0$ is the difference of the surface wave velocity in the measurement line v (with sensor structure) and in the reference line without coverage v_0.

In real implementations, the phase shift introduced by external elements (amplifier and transducers) is much less than the total phase shift of the line, i.e. $\phi_{ext} \ll \phi_0 = 2\pi n_0$ (typical values are: $\phi_{ext} \sim 80°$, $2\pi n_0 \sim 81{,}000°$, at $n_0 = 225$). Therefore, the difference in frequency compared to the reference line frequency is approximately equal to the relative speed change and the relative phase change of SAW:

$$\frac{\Delta f}{f_0} \approx \frac{\Delta v}{v_0} = -\frac{\Delta \phi}{\phi_0} \tag{4.3}$$

where $\Delta v = v - v_0$ is the difference of the surface wave velocity in the measurement line v (with sensor structure) and in the reference line without coverage v_0. The minus sign results from the fact that the phase of the wave is inversely proportional to its velocity.

In practice, the so-called *coverage factor* κ, which describes the fact that no sensor structure occupies the entire length of the L line, but always its part (Fig. 4.4) (typical values $\kappa = 0.7$–0.9). Therefore:

$$\kappa = \frac{L_s}{L} = \frac{n}{n_0} \tag{4.4}$$

The relative frequency changes obtained in practice are correspondingly smaller:

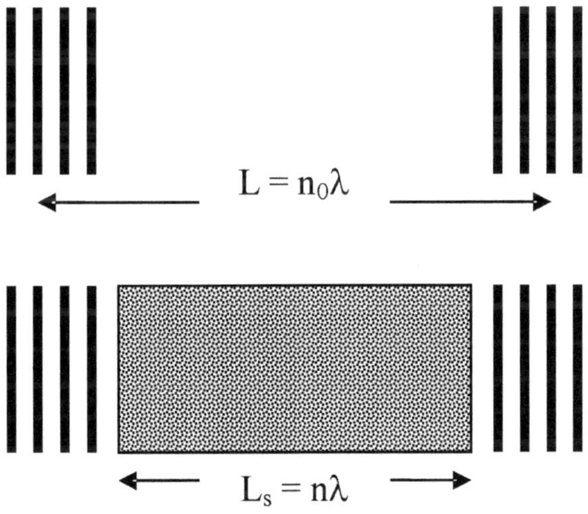

Fig. 4.4 Length of delay line L and length of structure L_s

$$\left(\frac{\Delta f}{f_0}\right)' \approx \kappa \frac{\Delta v}{v_0} \tag{4.5}$$

In the double-delay line measurement method, the relative change in the frequency difference of the two acoustic tracks (with sensor and reference structure) is approximately equal to the relative change in speed difference of the SAW propagating independently in each line, after taking into account the coverage factor κ.

In the case of testing sensor structures that give small changes in the wave velocity as a result of interaction with gases, the maximum value of this factor should be sought, which is the task of the masking process.

The double-delay line method in the form of measuring the differential frequency Δf is used to analyze the sensory properties of the structures made. We obtain information about how sensitive it is to a given type and concentration c of vapor or gas by observing the course of changes in the Δf signal. This signal increases or decreases when interaction is present, or remains unchanged in the absence thereof:

$$\frac{\Delta f(c)}{f_0} = \frac{f(c) - f_0}{f_0} \approx \kappa \frac{v(c) - v_0}{v_0} \tag{4.6}$$

A disadvantageous feature of the method is the tendency to jump changes in the differential frequency of the system, which results from the modal characteristics of the interdigital transducers in each of the lines [6]. During the tests, special attention should be paid to ensure that the obtained signal Δf under the established measuring conditions (temperature, flow) as a result of the metered concentration of the gas does not undergo jump changes. The method used in the author's research can be presented at several points.

1. Placing the two-channel sensor module in the measuring chamber equipped with the appropriate electronic system.
2. Oscillation excitation by selection of voltages powering the electronic system—this cause a change in the working points of the applied electronic elements.
3. Observation of the obtained signal Δf on the oscilloscope screen—correction of the supply voltages until a stable course is achieved.
4. Turning on the heating system and temperature stabilization.
5. Power supply voltage correction to get stable Δf signal at a given temperature.
6. Signal stabilization at specified values of structure temperature and carrier gas flow through the measuring chamber.
7. Metering of gases according to the assumed measurement cycle and acquisition of measurement results.

4.4 Method with Frequency Mode Identification

The resulting difference in frequency Δf results from the fact that the sensor structure is generated in the measurement line itself. Depending on the mechanical and electrical properties of the structures produced, the propagation velocity of SAW is reduced or increased. This leads to two possible configurations of the measurement frequency f and the reference f_0, this is the so-called normal and inverse configuration of frequency modes [6]. The normal configuration of frequency modes (**NMC**) occurs when $v < v_0$, i.e. $f < f_0$ (Fig. 4.5a), while the inverse (**IMC**) occurs when $v > v_0$, for $f > f_0$ (Fig. 4.5b).

The frequency mode identification method is a modification of the method with a double-delay line. It is used when we want to obtain additional information about the physical properties of the interactions, such as whether the SAW speed decreases or increases. One way is to ensure that the configuration of the frequency modes established at the beginning of the system does not change during the measurements. Another way is to monitor the frequency of each line or only the measured frequency throughout the duration of the measurements.

The mode identification method is used because, in practice, it is easier to measure a differential signal Δf, which has a much lower value of ~kHz than oscillation frequencies in the MHz range.

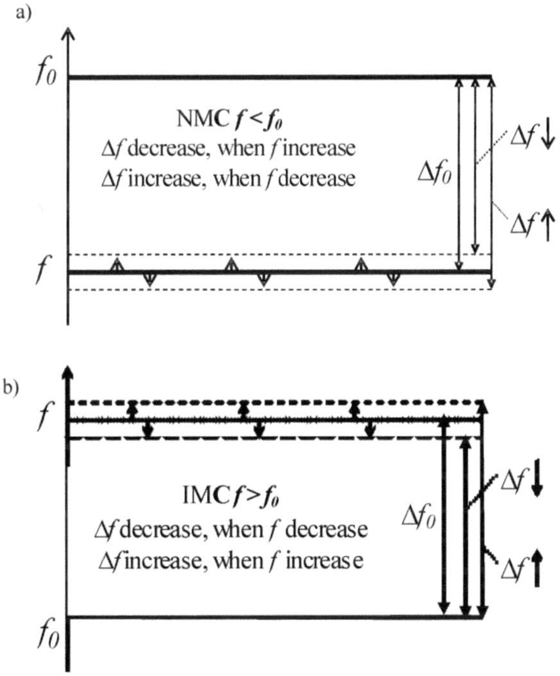

Fig. 4.5 Normal (**a**) and inverse (**b**) frequency mode configuration in a dual-delay line method

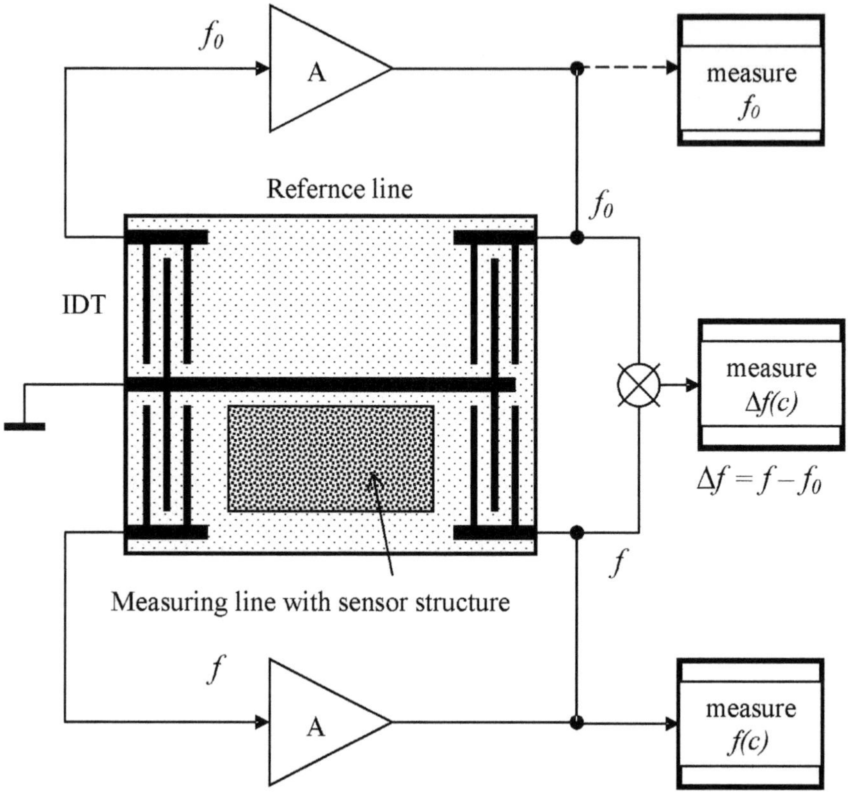

Fig. 4.6 The method with frequency mode identification

The scheme used by the author is extended by the possibility of measuring resonant frequencies f_0 and f of each line, which allows to identify frequency modes (Fig. 4.6).

The directions of changes in the speed of the SAW as a result of the interaction of the sensor structure with the gas molecules are determined by changes in its mechanical and electrical parameters. Each configuration of the frequency modes corresponds to different possibilities of perturbations of the surface density ρ_s, the elastic moduli (*ela.m*), and the electrical surface conductivity σ_s of the structure. The diagram for the NMC (Fig. 4.7) concerns a simple situation where only one parameter of the structure changes or changes in the other parameters interacts to increase or decrease the wave velocity. In the case of any combination of these changes, the outcome is difficult to predict. A similar scheme (Fig. 4.8) applies to IMC.

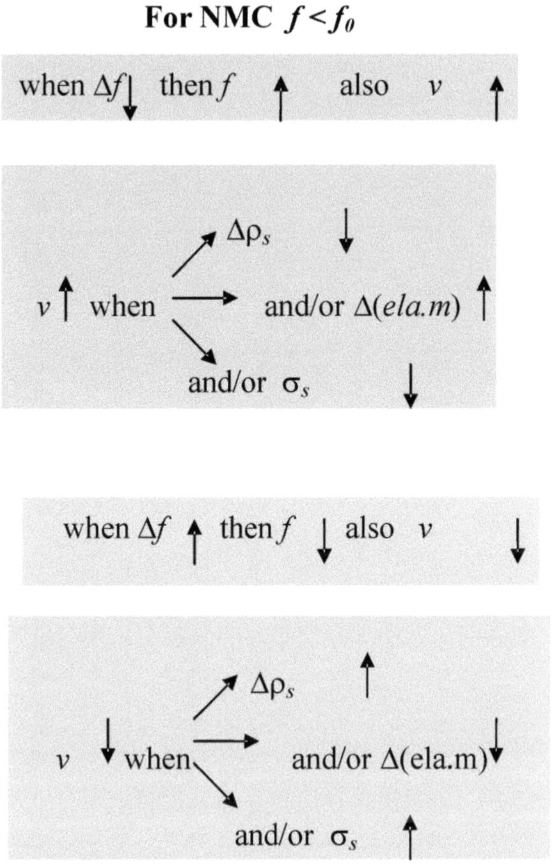

Fig. 4.7 SAW velocity changes interpretation for normal mode configuration (NMC), *ela.m*—means elastic modules of the structure

4.5 Method with Multi-channel Sensor Modules

Sensor matrices are commonly used to solve the selectivity problem of individual sensors. The author expanded the measuring system to four channels, which is an example of a matrix solution (Fig. 4.9). In this system, it is possible to study three different sensor structures and miniaturize the module (three active channels on a board of the same size). A different electronic system with three frequency mixers is necessary, resulting in three measurement signals (Fig. 4.10) [5]. Adding more channels increases the complexity of the electronic system, which causes many interferences, e.g. in frequency mixing.

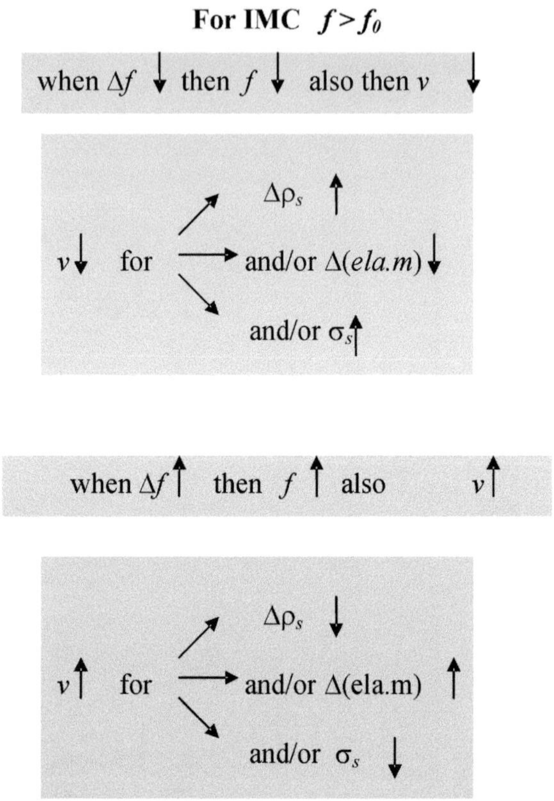

Fig. 4.8 SAW velocity changes interpretation for inverse mode configuration (IMC), *ela.m*—means elastic modules of the structure

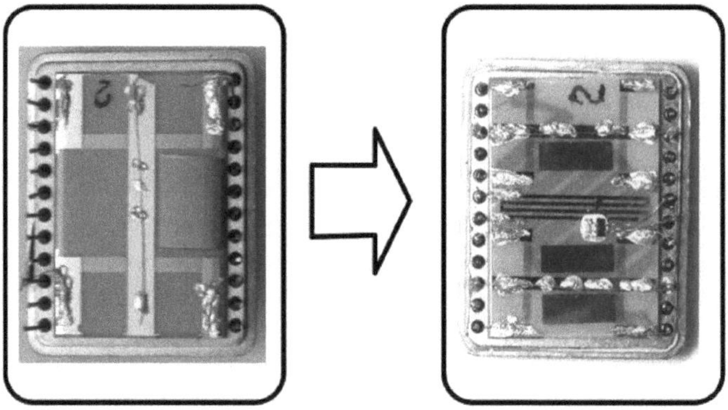

Fig. 4.9 Practical realization of the sensor module extension from 2 to 4 channels

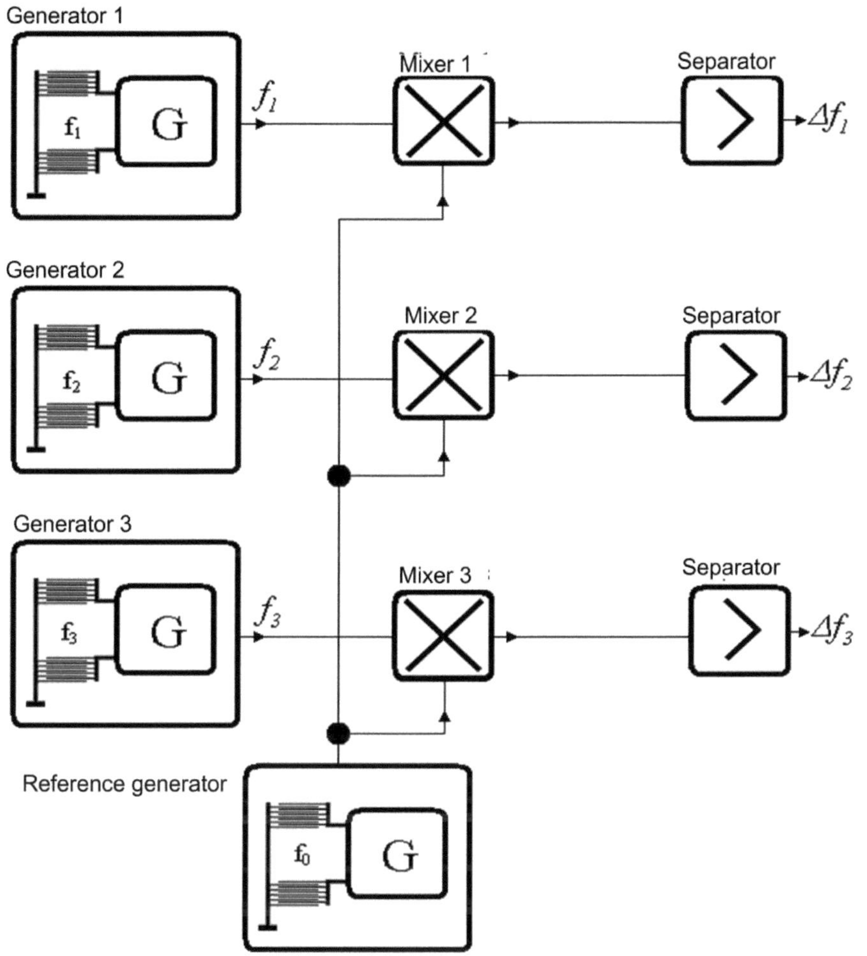

Fig. 4.10 Block diagram of the four-channel electronic system [7]

4.6 Acoustoelectrical Method

The acoustoelectrical method is cognitive in nature and is a combination of the acoustic method with the identification of frequency modes and an additional planar electrical method. Its essence is to carry out simultaneous acoustic and electrical tests for sensor modules prepared in the same technological processes, which ensures their almost identical structural structure. The method allows the correlation of the test results by comparing the obtained changes in speed of sensor structures (from the acoustic method) with changes in their electrical conductivity (from the direct electrical method) as a result of interaction with gas molecules. This allows us to determine whether the observed changes in velocity are the result of only the mass effect (no changes in structure electrical resistance) or possible participation of acoustoelectric interactions (there are then changes in sample resistance simultaneously with changes in frequency).

The planar electrical module consists of aluminum electrodes on a glass substrate on which the sensor structure is made. The electrode arrangement reduces the effective resistance of the test sample by connecting the particle resistances existing between each pair of electrodes. This ensures an increase in the intensity of the current measured at constant voltage, and more importantly, a decrease in the power emitted at constant current. Such a configuration makes it possible to study the changes in the electrical resistance of the produced sensor structures as a result of interaction with the gas molecules in their environment. At the measuring station used, the resistance changes were measured using the Agilent 34970A multiplexer for samples up to 100 MΩ and the Keithley 617 m for samples above this value (up to 200 GΩ).

For several phthalocyanine-palladium sensor structures, resistance measurements were made first for the single layer and later for the bilayer structure after the application of palladium. In all cases, a significant effect of palladium on reducing the resistance by several orders of magnitude was found. Detailed results of these studies are described in Chap. 5. The schematic view and SEM image of the electrical module structure together with the electrodes for testing by the planar electrical method is shown in Fig. 4.11. The surface conductivity of the investigated sensor structure can be estimated on the basis of the expression [8]:

$$\sigma_s = \frac{1}{R_{sam}} \frac{p}{(2N - 1)l} \tag{4.7}$$

where p is the electrode width and distance between electrodes ~50 μm, N is the number of electrode pairs = 9, l is the electrode coverage width ~10–14 mm depending on sample, R_{sam} is the sample resistance determined from measurements.

4.7 Measuring Station

Depending on the measurement method used, different elements of the universal measuring station built at the Institute of Physics of the Silesian University of Technology, currently the Department of Optoelectronics, are used. The station in the most developed version allows simultaneous measurement of three frequencies, two- or four-channel modules, placed in the respective measuring chambers. In addition, the electrical resistance of the structure is monitored in the acoustoelectrical method. The heating system allows increasing and stabilization the temperature on the surface of the sensor modules to approx. 120 °C (±0.1 °C).

The general scheme of measurement setup for the sensor structures tested by the methods discussed is shown in Fig. 4.12. Four basic systems can be distinguished here:

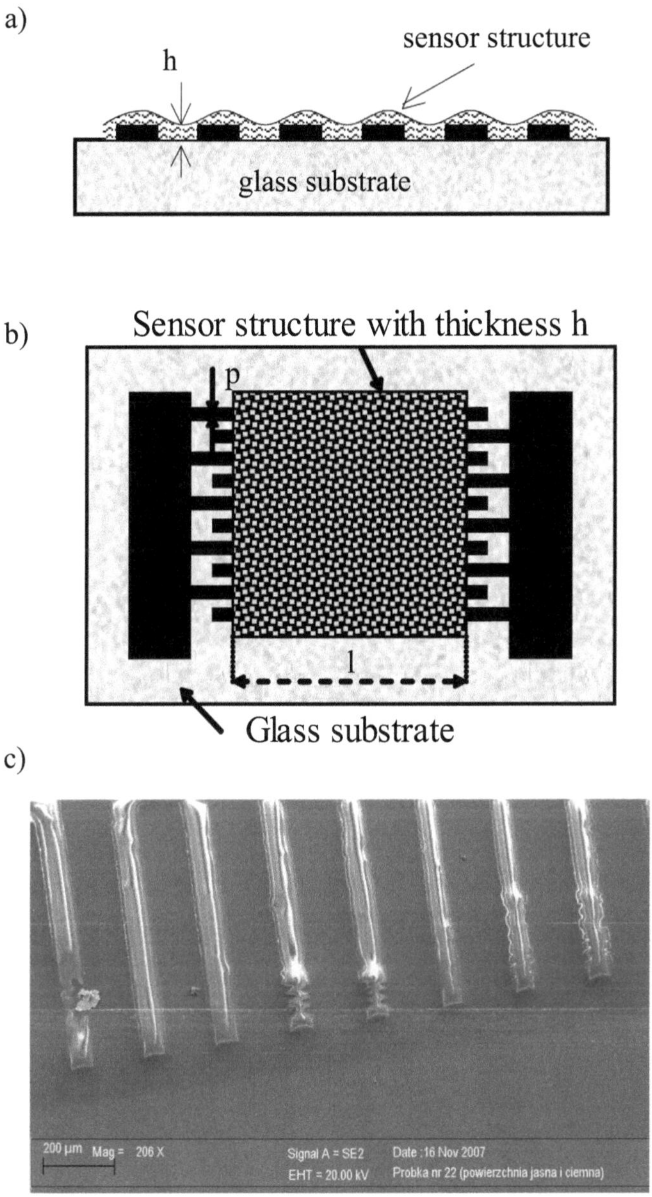

Fig. 4.11 Electric planar module: **a** view along electrodes **b** top view **c** SEM picture of the bilayer structure 160 nm H_2Pc + 27 nm Pd on a glass substrate

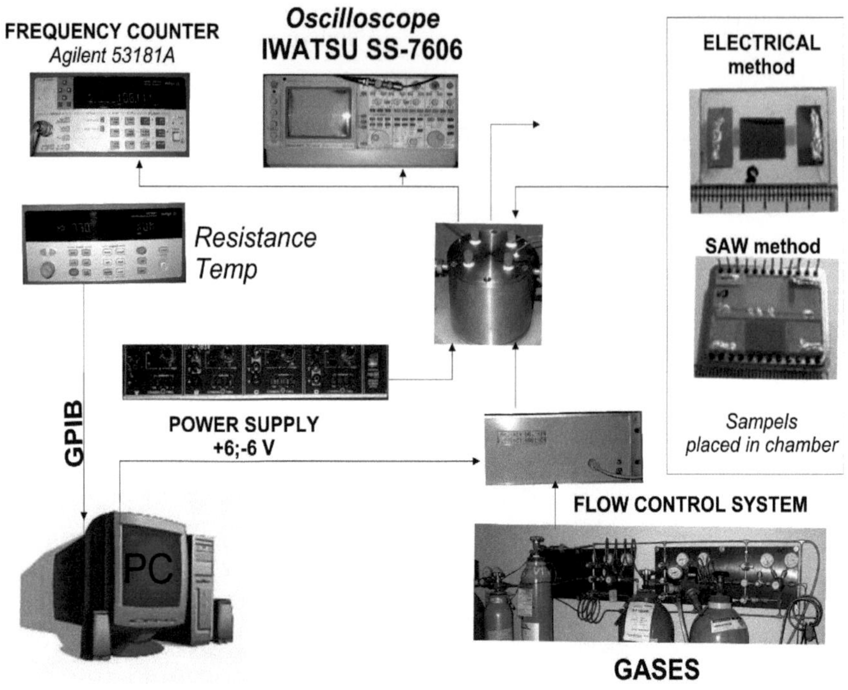

Fig. 4.12 General scheme of the experimental setup

(a) measuring chamber with electronic system and sensor modules—the proper operation of the electronic system depends on its distance to the acoustic module, the smaller the distances, the less interference of the system and easier excitation of oscillations,
(b) heating and temperature stabilization system—extremely important due to the strong dependence of the frequency signal on temperature fluctuations,
(c) system of measurement, observation, and acquisition of results—monitoring of differential frequency and/or oscillation frequencies of each track, monitoring of resistance of electrical structures, observation of frequency signals, acquisition of test results through GPIB bus,
(d) metering system—production of different concentrations of gases in the measuring chamber by means of standard mixtures (Air Product, Linde, Degas) at a fixed total flow, additionally changing the humidity of the atmosphere through the water vapor dispenser together with the sensor.

In the acoustic-electrical method, acoustic and electrical modules are placed close together in a measuring chamber (Fig. 4.13). The temperature of each module is measured with a thermocouple or Pt100 element attached to their surface.

Fig. 4.13 Measurement chamber prepared for the acoustoelectrical method

4.7.1 Chambers and Measuring Instruments

The measuring chamber (except with multi-channel modules) is shown in Fig. 4.13. It is made of stainless steel in the shape of a cylindrical container and a closing cover, which also serves as a supporting element for the electrical module in the acoustoelectrical method.

In the lower part of the chamber are the corresponding electronic circuits. The chamber allows testing the properties of interchangeable sensor modules (acoustic and electrical) at different interaction temperatures, which are increased by a thick-layer resistor made on an almond (alund). The surface temperature of each module placed in the chamber is precisely controlled by a miniature J-type thermocouple or a Pt100 element attached to their surface.

Digital Frequency Meters and Multiplexer—Agilent's 53131A, 53181A, and 34970A digital frequency meters are used to monitor the temperature and resistance of electrical module structures up to 100 MΩ. The differential frequency as well as the oscillation frequency of individual acoustic lines can be measured.

Digital oscilloscope allows observation of frequency signals (Δf) and delay line oscillation signals (f and f_0). Iwatsu's SS-76S06 or Agilent's DSO3202A oscilloscope was used.

The Keithley 614 electrometer is used to measure the resistance of sensor structures when their total resistance exceeds 100 MΩ. It has a large measuring range of up to 200 GΩ.

Control and measurement controller was used to read and record the frequency, temperature and resistance of the tested sensor modules with a set time step and save the results in a text file. It enables visualization of recorded runs during research and archiving of selected research results through the GPIB data bus. It was designed and executed by a team as part of one of the research projects.

4.7.2 Gas Metering System

The metering system is used to set a specific gas mixture with the desired concentration of the tested gas in synthetic air or nitrogen (carrier gas). The gas flow is controlled by means of precise mass flow meters installed in the individual gas supply tracks. The system allows mixing of three different gases with the carrier gas and controlling the humidity of the atmosphere.

The toxic gases used in the studies are sample mixtures of a specific gas in the synthetic air. A metering system is used to obtain the expected gas concentrations, in which the sample mixtures are combined with other gases and synthetic air, is used. A mass controller with a measurement range of up to 1000 ml/min was installed in the carrier gas track. In the other two tracks there are similar devices for 100 ml/min and in the fourth for 500 ml/min. The flow is switched on and off automatically by pneumatic valves.

Due to the fact that the process of filling the measuring chamber and feeding tubes takes some time, the dynamics of the filling process of the measuring chamber in order to achieve a constant gas concentration need to be explained. The equation describing this phenomenon has the following form [2]:

$$\frac{dc}{dt} + \frac{Q}{V_0}c(t) = \frac{Q}{V_0}c_0 \tag{4.8}$$

where c_0 is the gas concentration in the chamber at steady state, Q is the gas flow, and V_0 is the volume of the measuring chamber.

Solving the above equation, we will get an expression defining the change in time concentration of the gas in the measuring chamber during its filling:

$$c(t) = c_0\left[1 - \exp\left(-\frac{Q}{V_0}t\right)\right] \tag{4.9}$$

The time constant of the chamber filling process is

$$\tau_K = \frac{V_0}{Q} \tag{4.10}$$

It is assumed that the gas concentration in the chamber is determined after a time equal to 5 τ_K [3]. The time to reach a fixed gas concentration in the chamber for the chamber together with feeds with a volume of about 200 ml at a flow rate of 1000 ml/min is ~1 min. In the case of a system with four-channel sensor modules, the volume of the chamber is limited to 6 ml, which gives a filling time of ~2 s.

4.8 Phenomena Related to the Methodology of Testing Sensors with SAW

Many years of measuring practice of the author allows indicating the most important phenomena related to the methodology of testing gas sensors with SAW. These include mainly the phenomenon of the influence of long-term thermal drift on the frequency stability of oscillators and possible to obtain frequency change without unfavorable jumps to other oscillatory modes.

4.8.1 Thermal Drift

The long-term thermal drift is mainly caused by the temperature instability of the piezoelectric substrate along with the sensor structure made on its surface. Depending on the frequency mode configuration (NMC or IMC) of the exemplary sensor structure, both increasing and decreasing the differential frequency can be obtained. Furthermore, the direction of the changes is dependent on whether the temperature increases or decreases, therefore, depending on when the sensor is switched on, both effects can be observed. This is due to the fact that the temperature may initially decrease as a result of switching on the gas flow, and then gradually increase as a result of the heat emitted from the electronic circuits. After a sufficiently long time (1–2 h) the differential signal is stabilized. The time to obtain stable differential signals is also dependent on the sensor structures used. Figure 4.14 shows an example of a single 20 nm Pd sensor structure, performed on $LiNbO_3$ Y–Z, illustrating the phenomenon of thermal drift.

As a result of a constant flow of gas of 1000 ml/min (air and two concentrations of hydrogen 1.5 and 2%) through the measuring chamber, a systematic reduction of the differential signal is observed. This is associated with an increase in the temperature of the sensor structure by about 0.1 °C.

In laboratory practice, depending on the tested structure and the piezoelectric substrate used, a sufficiently long initial stage of metering of the carrier gas with a fixed flow should be taken into account, which makes it possible to minimize the thermal drift. However, if the tests were carried out in an unstabilized system, the solution is a graphical correction of the resultant course. The thermal drift should be estimated (in short time ranges it will usually be a linear function, in longer time ranges—square) and then corrected the frequency signal Δf.

An example of positive thermal drift and its correction for a bilayer sensor structure is shown in Fig. 4.15.

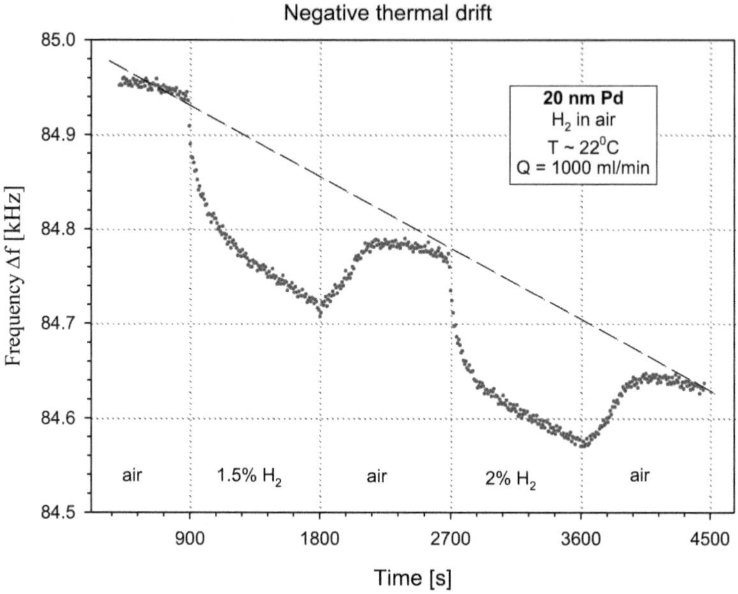

Fig. 4.14 Negative thermal drift for single sensor structure of 20 nm of palladium (Pd) on $LiNbO_3$ Y–Z substrate

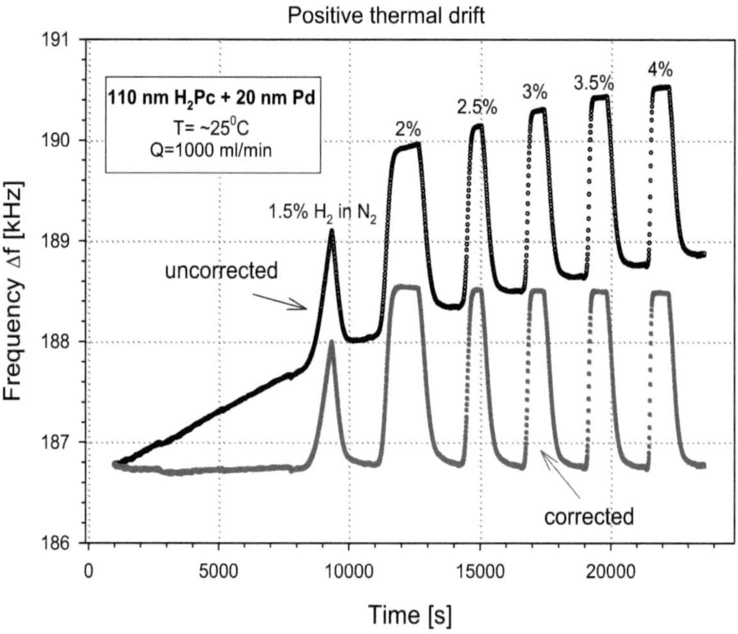

Fig. 4.15 Positive thermal drift for the bilayer sensor structure (metal-free phthalocyanine + palladium) with its correction—interaction with hydrogen in nitrogen

4.8.2 Frequency Changes

The measure of the interaction of the studied sensor structure with the gas molecules in its environment is the change in the differential frequency, resulting from changes in the measuring frequency, which in turn is the result of changes in the speed of the SAW. The greater the interaction is, the greater the change in frequency and the possibility of practical use of the structure. It should be noted, however, that sensor modules with SAW are generally intended for small and possibly medium concentration ranges of various types of gases. Large concentrations lead to significant frequency shifts, which can cause unfavorable jumps of the system to other oscillation modes. In practice, frequency shifts were observed at the level of 10 kHz and even 20 kHz (for layer structure type WO_3 + Pd at exposure temperature increased to approx. 100 °C and 2% hydrogen concentration in dry air RH~5%), without adverse phenomena associated with frequency mode jumps. However, for such large shifts, the frequency signal is unstable, which is particularly evident for the structure 150 nm WO_3 + 10 nm Pd (Fig. 4.16 green curve).

This is the result of testing in a prototype four-channel matrix system (three active channels and one reference channel) that allows the simultaneous testing of

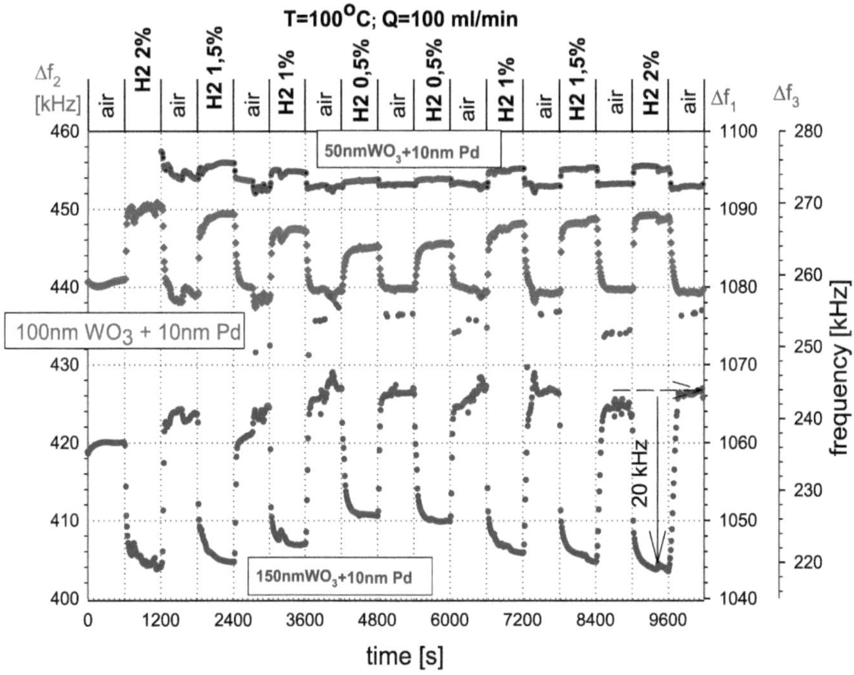

Fig. 4.16 An example of a very high frequency shift (~20 kHz) with unstable signal in Δf_3 channel for 2% concentration of hydrogen in air and layered structure 150 nm WO_3 + 10 nm Pd at 100 °C on $LiNbO_3$ Y–Z substrate; gas flow Q = 100 ml/min

three different structures under identical measurement conditions. The effect of the tungsten oxide (WO_3) layer thickness is clearly visible because the palladium layer had the same 10 nm thickness. The third channel (Δf_3), which had a structure 150 nm WO_3 + 10 nm Pd, operated in an inverse mode configuration than the Δf_1 and Δf_2 channels, because each concentration of hydrogen corresponded to a reduction in the frequency signal Δf_3 (green), the opposite of that for the other channels.

4.9 Summary

This chapter discusses the most important measurement methods used in SAW gas sensors.

The problem that needs to be paid special attention to is the precise stabilization of the temperature on the crystal surface in a large time interval. The aim is to obtain satisfactory stability of the frequency signal, because the compensation in the differential method of two delay lines affects the substrate, while the influence from the side of the sensor structure is no longer compensated in this way. The signal stability is related to the high temperature coefficient of the piezoelectric substrates used, e.g. $LiNbO_3$ Y–Z (~−94 ppm/°C), which causes a change in the wave propagation velocity. So-called thermally compensated substrates with zero velocity temperature coefficient (e.g. ST-X quartz) can be used, but they have a much lower electromechanical coupling coefficient, which drastically reduces the usability of acoustoelectric interactions. For semiconductor–metal bilayer structures, piezoelectric substrates with a high electromechanical coupling coefficient such as lithium niobate Y–Z ($K^2 = 4.8\%$) are best suited. Temperature stabilization, using the Pt100 element and the Shimaden SR-94 temperature controller, is achieved at about ±0.1 °C. In previous studies, miniature J-type thermocouples were used, but they provided slightly worse temperature stabilization of ±0.3 °C, causing the observed temperature drift.

Temperature drift is a characteristic phenomenon in sensor module studies with SAW. This effect can be corrected by improving temperature stabilization. Commercial biosensors with SAW (e.g. Nanofilm) use two-stage temperature stabilization, even at 0.001 °C.

References

1. Frye GC, Martin SJ (1993) Velocity and attenuation effects in acoustic wave chemical sensors. In: Ultrasonic symposium, vol 1. New York, pp 379–383
2. Jakubik W (1997) Propagacja akustycznej fali powierzchniowej w cienkich warstwach wybranych ftalocyjanin (Propagation of surface acoustic wave in thin films of selected phthalocyanines). Dissertation, Silesian University of Technology, Gliwice
3. Nieuwenhuizen M, Nederlof A, Barendsz W (1988) Metallopthalocyanines as chemical interfaces on a surface acoustic wave gas sensor for nitrogen dioxide. Anal Chem 60(3):233–239

4. Ricco AJ, Martin SJ, Zipperian TE (1985) Surface acoustic wave gas sensor based on film conductivity changes. Sens Actuators 8:319–333
5. Setkiewicz M (2008) Opracowanie układu rejestracji wielokanałowego systemu z AFP (Development of a multichannel recording system with SAW). Master thesis Silesian University of Technology, Gliwice
6. Urbańczyk M, Jakubik W (2000) Studies of the electrical and mass effects in surface acoustic wave gas sensor. Electron Technol 33(1):161–166
7. Urbańczyk M, Jakubik W, Kochowski S (1994) Investigation of sensor properties of CuPc with the use of SAW. Sens Actuators B 22(2):133–137
8. Wohltjen H, Barger W, Snow A, Jarvis NL (1985) A vapour-sensitive chemoresistor fabricated with planar microelectrodes and a Langmiur-Blodgett organic semiconductor film. IEEE Trans ED 32(7):1170–1174
9. Wohltjen H, Dessy R (1979) Surface acoustic wave probe for chemical analysis. Anal Chem 51:458–1470

Chapter 5
Study of Selected Sensor Structures with SAW

Abstract This chapter describes the most important experimental work carried out by the author using sensor modules with double-delay line, made on the substrate of Y–Z lithium niobate. This substrate is characterized by a good electromechanical coefficient, which provides in many cases decent acoustoelectric responses. The work concerns the sensory properties of selected single, bilayer structures, as well as prototype multi-layer structures, in which additional layers perform limiting functions, e.g. the influence of water vapor. However, the most attention has been paid to the semiconductor–metal bilayer structures, e.g. phthalocyanine-palladium and tungsten oxide-palladium, as they are the most promising in terms of practical applications in hydrogen sensors.

In order to present the significant benefits of using various sensor structures in gas sensors with SAW will be presented in order:

- test results of sensor properties against hydrogen for single layers of palladium, phthalocyanines, and tungsten oxide (WO_3)
- results of studies of selected bilayer structures of three main types: MPc-Pd (phthalocyanine + palladium), MO-Pd (metal oxide + palladium) and for the structure Nafion-PANI (nafion + polyaniline)
- test results of several layer structures with prototype polyethylene (PE) and polymethylopentane (TPX) protective layers.

The research was carried out using the discussed measurement methods: methods with double-delay line (also with identification of frequency modes) and acoustoelectric method, enabling correlations of the obtained responses. The latest test results were made using a four-channel sensor module, which allows the simultaneous testing of three different sensor structures. Due to the thin layers of palladium used (in the vast majority of cases), the researchers focused on testing hydrogen up to the limit of explosiveness in the air, i.e. to a concentration of 4% by volume. They also tested for lower concentrations with other gases, i.e. nitrogen dioxide, carbon monoxide, ammonia, or methane.

W. Jakubik, *Application of Surface Acoustic Waves in Gas Sensors*, Springer Series in Surface Sciences 69, https://doi.org/10.1007/978-3-031-89971-3_5

5.1 Single Sensor Structures

5.1.1 Palladium Layers

The palladium (Pd) thin layers were applied to a piezoelectric substrate with $LiNbO_3$ Y–Z in the process of vacuum evaporation from a tungsten heater. The rate of application was approximately 0.3–0.5 nm/s. The resulting thicknesses are in the range of 10–30 nm.

The temperature drift-adjusted result of the interaction of a single layer of palladium with hydrogen in the air is shown in Fig. 5.1, which is derived from Fig. 4.14. Small changes in the frequency of ~120 Hz, for both concentrations of hydrogen (1.5 and 2%), indicate mass interactions, which are very small for thin layers and extremely light hydrogen ions. Mass interactions can be further confirmed by a method with frequency mode identification. In the case of Fig. 5.1, we are dealing with an inversion configuration (IMC). The frequency of the oscillator loop, in which the palladium layer f_{Pd} ~ 43,614 kHz is located, is greater than the frequency of the reference line f_0 ~ 43,529 kHz.

As a result of the interaction of hydrogen with the palladium layer, a decrease in the differential frequency Δf was observed, and a decrease in the frequency of f_{Pd} in the measurement path with the layer, which in turn is the result of a decrease in the propagation velocity of the SAW. According to the diagram of changes in the SAW

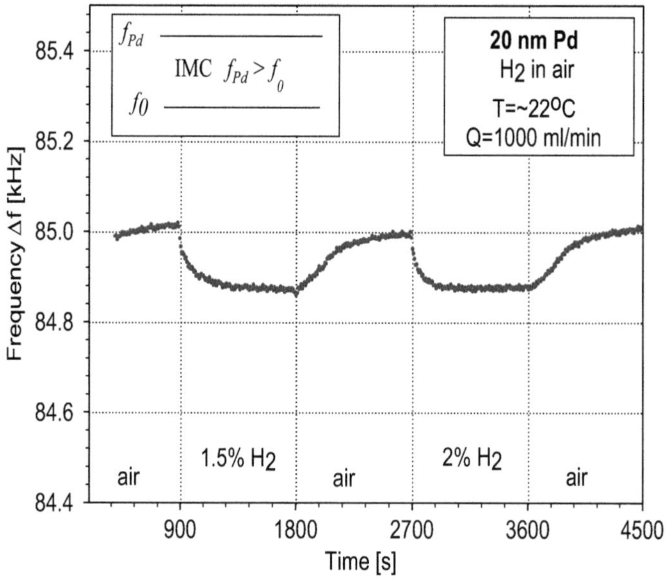

Fig. 5.1 The result of mass interactions of a single 20 nm thick palladium layer at a temperature of ~22 °C with hydrogen of 1.5 and 2% concentrations in the air [18]

velocity for the IMC (Fig. 4.8), its decrease is possible as a result of mass-elastic interactions and/or acoustoelectric interactions, but then the surface conductivity of the palladium layer σ_s as a result of interaction with hydrogen would have to increase. However, the observations of changes in this conductivity for samples of thin palladium layers were inverse. As a result of interaction with hydrogen, the resistance of Pd thin layers increases [4], and thus their surface conductivity decreases. The observed variations in the differential frequency of a thin single palladium layer are therefore due to mass-elastic interactions.

Using the presented model for mass-elastic interactions, perfectly elastic structures (3.27), it is possible to perform their quantitative analysis.

5.1.2 Analysis of Mass-Elastic Interactions for Pd Film

The observed reduction in differential frequency (Fig. 5.1) as a result of the interaction of the Pd thin layer with hydrogen (1.5% or 2%) results for IMC from a reduction in the speed of SAW propagation. In the case of interactions of mass-elastic sensor structures with gases, this reduction in SAW speed is possible (see Fig. 4.8), if:

- the surface density of the structure increases and/or
- the elastic modules of the structure are decreasing (shrinking).

The relative changes in the differential frequency for this type of interaction can be recorded as follows:

$$\delta\left(\frac{\Delta f}{f_0}\right)=\kappa\left(\frac{\delta(\Delta v)}{v_o}\right)_{me}\underbrace{\approx -c_{m_1}\kappa f_o(\delta\rho_s)}_{S_1}+\underbrace{c_{m_2}\kappa f_o h\left(\delta(\mu'\frac{\lambda'+\mu'}{\lambda'+2\mu'})\right)}_{S_2} \tag{5.1}$$

where $\delta\left(\frac{\Delta f}{f_0}\right)=\frac{\Delta f\,(\text{hydrogen})-\Delta f\,(\text{air})}{f_0}=\frac{\sim 84880-85000}{43.6\times10^6}\approx -2.75\times10^{-6}$

c_{m1}—mechanical coupling constant dependent on the substrate ~5.5×10^{-8} m^2s/kg for $LiNbO_3$ Y–Z ("pure mass effect"),

κ—structure coverage factor ~0.8 for the test sample,

$\delta\rho_s$—changing the surface density of the palladium layer as a result of interaction with hydrogen,

c_{m2}—fixed couplings of "stiffening" surface ~5.7×10^{-15} s^3/kg for $LiNbO_3$ Y–Z,

f_0—frequency of wave propagation on free crystal surface ~43.6 MHz,

h—layer thickness ~20 nm,

$\left(\delta\left(\mu'\frac{\lambda'+\mu'}{\lambda'+2\mu'}\right)\right)$—change in the elastic modulus of the palladium layer as a result of interaction with hydrogen, in short denoted as $\delta(ela.m)$,

λ', μ'—Lame constants for Pd thin layer.

The S_1 component of the "pure mass effect" resulting from an increase in the surface density of the layer by reaction with hydrogen can be estimated as follows:

$$S_1 = -c_{m_1}\kappa f_o(\delta\rho_s) \tag{5.2}$$

where $\delta\rho_s = \frac{\delta m}{S} = N_s m_H$ is the change in the surface density of the layer, resulting from the average density of palladium atoms on its surface $N_s \sim 2 \times 10^{19}$ m^{-2} and the mass of a single hydrogen atom $m_H \sim 1.67 \times 10^{-27}$ kg. Assuming total coverage of all active surface centers (the saturation state of the course $\Delta f(t)$ is achieved practically after 300 s for concentrations of 1.5 and 2% hydrogen in air) and the absence of volume absorption, the value of component S_1 is much smaller than the observed change in differential frequency and is

$$S_1 \cong -0.064 \times 10^{-6} < \delta\left(\frac{\Delta f}{f_0}\right) \cong -2.75 \times 10^{-6} \tag{5.3}$$

The estimated value of component S_1 is too small for typical values of the surface density of palladium atoms. Assuming its greater value, e.g. $N_s \sim 10^{21}$ m^{-2} (which is strongly dependent on the structural structure of the resulting layer), it is possible to achieve conformity at the level of the same orders of magnitude. The calculated value of component S_1 is then approximately equal to the measured relative change in frequency Δf:

$$S_1 \cong -3.2 \times 10^{-6} \cong \delta\left(\frac{\Delta f}{f_0}\right) \cong -2.75 \times 10^{-6} \tag{5.4}$$

The observed even small changes in the differential frequency for the thin layer of palladium interacting with hydrogen can also be explained by the second component S_2 of the dependence (5.1):

$$S_2 = c_{m_2}\kappa f_o h\left(\delta\left(\mu'\frac{\lambda' + \mu'}{\lambda' + 2\mu'}\right)\right) \tag{5.5}$$

This component is associated with the elastic parameters of the sensor structure. As already mentioned, its decrease causes a decrease in the speed of SAW propagation and vice versa. The observed decrease in SAW speed may result from a decrease in the factor associated with the elastic modules of the layer. Such effects have been observed in studies by other authors who have found that the elastic modules of Pd thin layers undergo significant changes as a result of interaction with hydrogen [26]. Table 3.1 presents the results of studies by Anisimkin [1] which show that the elastic modules can decrease as well as increase under the influence of hydrogen absorption by the palladium layer. However, the functional form of component S_2 indicates that changes in the μ' palladium modulus, which usually decreases, are more important. However, the estimated value of component S_2, assuming a 29% decrease in μ' (see Table 3.1), differs by one order of magnitude from the observed relative frequency changes.

$$S_2 \cong -27.8 \times 10^{-6} > \delta\left(\frac{\Delta f}{f_0}\right) \cong -2.75 \times 10^{-6} \quad (5.6)$$

It follows that the explanation of the observed changes only by the S_2 effect (dependent on the elastic modulus of the layer) requires an even greater reduction of the modulus of the thin palladium layer. In most cases, the random frequency changes resulting from the interaction of Pd layers with hydrogen will be a combination of both the S_1 and S_2 effects. The mass-elastic interactions of thin palladium layers with hydrogen will result both from an increase in their surface density and from a decrease in their elastic modulus.

The small changes in frequency, even for relatively large concentrations of hydrogen, despite the use of a very good hydrogen-absorbing material such as palladium, show that the single, metallic sensor structures in the SAW system do not give satisfactory results. The sensor response is the result of small mass or mass-elastic interactions. Much thicker Pd layers should be used, allowing hydrogen to diffuse into the volume, but then the reaction times are significantly prolonged (~several minutes). Such Pd structures are unsuitable for sensor applications [2].

5.1.3 Copper Phthalocyanine Layers (CuPc)

The phthalocyanine layers were made by vacuum evaporation on a $LiNbO_3$ substrate for the acoustic module and in some cases on a glass substrate of electrical modules. The technological parameters of their execution were similar [11]: vacuum 6–8 × 10^{-6} hPa, steam source temperature: 320, 345, and 370 °C, substrate distance from steam source approx. 80 mm. Before starting the proper evaporation process, phthalocyanins were degassed at ~200 °C for approx. 20 min. After evaporation at a given temperature and at the appropriate evaporation time, the source was cooled to ambient temperature for approx. 30 min, after which the sample was removed. By changing the evaporation time, it was possible to obtain layers of different thickness at the same temperature. The evaporation of the evaporation was controlled by a quartz resonator, which was scaled using optical interference method.

The result of the interaction of the ~720 nm CuPc layer with a 3% concentration of hydrogen mixed with nitrogen is shown in Fig. 5.2. The difference frequency increase of 350 Hz occurs only at a temperature increased to ~70 °C. A delay in the response of the structure to the high concentration of hydrogen is visible, which may be due to its free diffusion into the volume of a relatively thick layer. The pulses during the on/off dosing of hydrogen are due to the high sensitivity of the system to temperature disturbances.

The normal frequency mode configuration (NMC $f_{\mathrm{CuPc}} < f_0$) indicates that as the CuPc layer interacts with hydrogen, the measured frequency of f_{CuPc} decreases, and thus the speed of SAW propagation also decreases. According to the scheme of possible interactions (see Fig. 4.7) (as for a single palladium layer), the surface density of the layer may increase, and its elastic modulus may also decrease. In

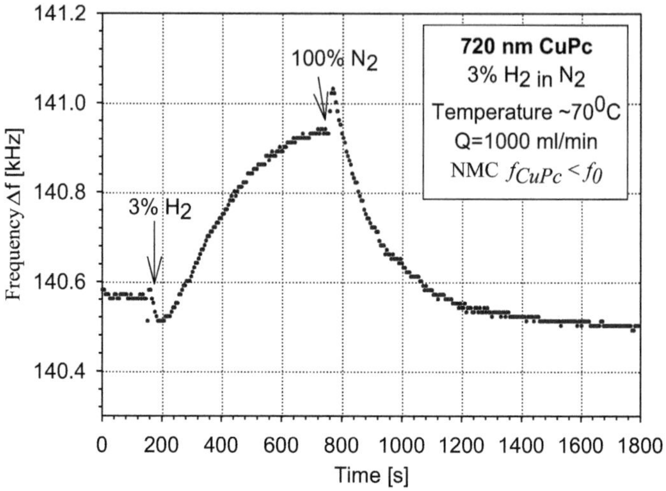

Fig. 5.2 Response of single layer of CuPc with thickness ~720 nm at ~70 °C for 3% hydrogen in mixture with nitrogen [11]

addition, if the surface electrical conductivity of the CuPc layer increased, this would allow for the co-operation of acoustoelectric interactions with the mass-elastic ones. However, the observed small changes in frequency (~350 Hz) indicate that this type of interaction is not occurring. This is confirmed by literature reports, where for most of the tested layers of phthalocyanines (including CuPc) made by similar technologies, a reduction in surface conductivity was registered in the case of interaction with hydrogen [28]. Such interactions, depending on the acoustoelectric parameter of a given sensor layer, reduce the SAW velocity, while reducing the measuring frequency changes. In some cases, acoustoelectric interactions may not occur at all, since there is no change in the electrical conductivity of the sample as a result of the interaction with the gas. An example of this type occurs for thin layers of cobalt phthalocyanine (CoPc) interacting with hydrogen, which were tested using the acoustoelectrical method [22].

5.1.4 Tungsten Oxide (WO_3) Layers

The WO_3 layers were obtained by vacuum evaporation at ~900 °C from commercially available powders (Aldrich, Sigma). Evaporation rates of about 1 nm/s were obtained. The interaction of these layers with hydrogen in air at ~34 °C was presented in paper [19] Fig. 4.

There is no change in both the differential frequency and the resistance of the respective structures. Like single thin layers of palladium and phthalocyanine, WO_3 thin layers do not show satisfactory results for hydrogen detection.

Selected single sensor structures, i.e. metallic layers of palladium and semiconductor layers (organic and inorganic), do not change their material parameters so much as a result of interaction with hydrogen as to be important for the disruption of SAW propagation. Only the copper phthalocyanine layer shows a certain tendency for stronger hydrogen sorption. Increasing the temperature of the interaction only slightly improves the magnitude of the frequency changes obtained, which for the CuPc layer are at a maximum level of ~350 Hz at a temperature ~70 °C.

5.1.5 Single Layers in NO_2 Detection

Single layers are used in the detection of gases other than hydrogen. The use of phthalocyanine layers is particularly effective for detecting trace amounts of NO_2 [5] and vapor of organic solvents in the air [6].

In terms of detecting low concentrations (~ppb) of NO_2 in the air, single layers of lead phthalocyanin (PbPc) show very good results. The frequency changes obtained, depending on the concentration and temperature of the impact, are several hundred times greater than in the detection of hydrogen. One of the many results obtained by the author is shown in Fig. 5.3 [15].

The differential frequency change of the sensor module with SAW is even about 3 kHz as a result of the presence of a very low concentration of NO_2 ~640 ppb in

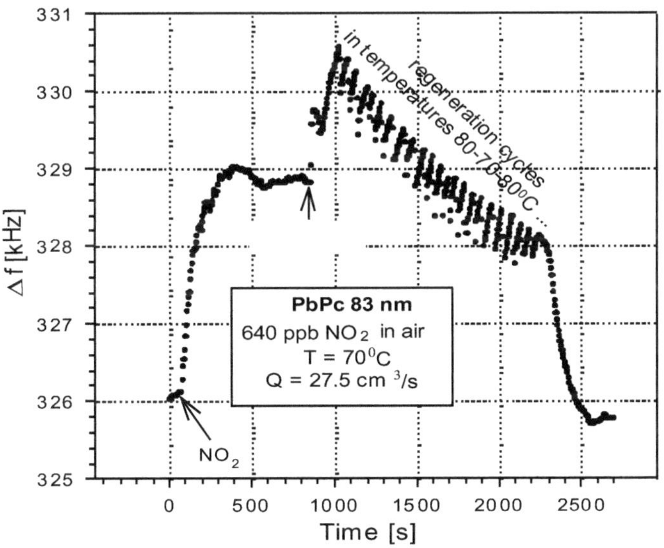

Fig. 5.3 Response of single layer of lead phthalocyanine (PbPc) with thickness ~83 nm at temperature ~70 °C against NO_2 at ~640 ppb in air with multiple regeneration cycles in temperatures 80–70–80 °C [15]

the air, at an exposure temperature of ~70 °C. Such large frequency changes indicate an extremely strong interaction of the PbPc layer with NO_2 molecules. The problem then becomes the time of spontaneous regeneration of the layer, which lasts several or even several hours with additionally reduced pressure in the measuring chamber, facilitating the desorption of NO_2.

In the same paper [15], a method of accelerated regeneration of the PbPc layer was presented. The structure was subjected to multiple heating and cooling cycles in the range of 80–70–80 °C, etc. In this way, the complete regeneration of the layer was achieved after ~1500 s, for an air flow ~27.5 cm^3/s through the measuring chamber (Fig. 5.3). For comparison, the time of complete regeneration of such a layer without using such cycles is from several to several dozen days.

5.2 Bilayer Structures

Due to the small changes in the frequency of single sensor structures, hydrogen detection is difficult in SAW sensors. Even the use of palladium, the material that best absorbs hydrogen, does not yield satisfactory results. Small mass-elastic interactions are observed, associated with increased mass and disruption of the elastic modules of the palladium layer.

The original solution, which significantly increases the sensitivity to hydrogen, is the use of bilayer sensor structures. Making such structures from materials that as single layers do not show significant interactions with hydrogen, e.g. phthalocyanine and WO_3 layers covered with thin layers of palladium, proves that this solution is extremely effective. The most important results of the study of selected bilayer structures of phthalocyanine-palladium and metal oxide-palladium have been presented.

5.2.1 Phthalocyanine-Palladium Bilayer Structures

5.2.1.1 Copper Phthalocyanine-Palladium Structures

Atomic force microscopy (AFM) images of bilayer structures, composed of a copper phthalocyanine layer of ~80 nm thickness and a palladium layer of ~18 nm thickness, are presented in Fig. 5.4. The structures were applied to $LiNbO_3$ substrate for acoustic method and to glass for electrical method in a single technological process. The resulting surface images of these structures (on $LiNbO_3$, and on glass) are very similar.

The use of the acoustoelectrical method allows simultaneous monitoring of the interactions for the acoustic and electrical module (Fig. 5.5). For the tested bilayer structure with a thickness of CuPc (~80 nm) at a temperature slightly increased to ~33 °C, the maximum differential frequency changes are about 500 Hz, for the

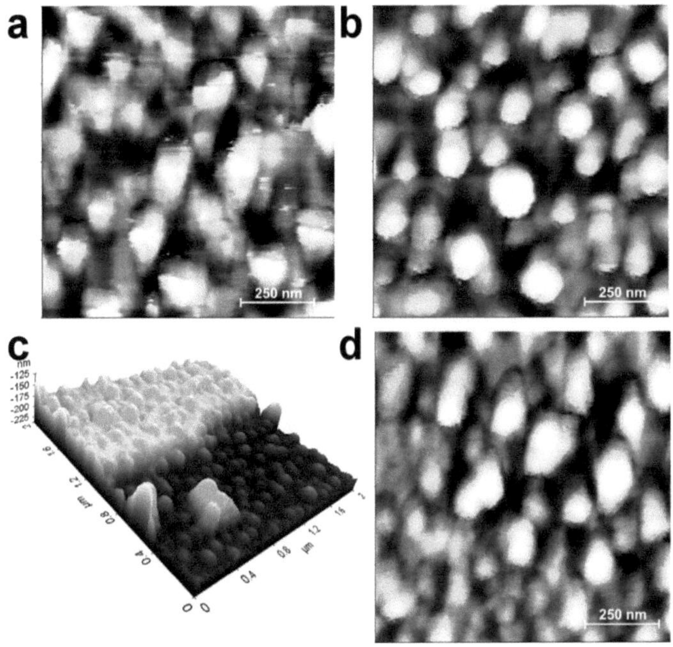

Fig. 5.4 AFM pictures of the investigated bilayer structures CuPc 80 nm + Pd 18 nm: **a** on $LiNbO_3$ Y–Z (SAW module) 1 × 1 µm; **b** on glass substrates between electrodes (electrical module) 1 × 1 µm; **c** on glass substrate—aluminum electrode edge 2 × 2 µm in 3D; **d** on glass substrate on an electrode (left bottom corner) 1 × 1 µm

highest concentration of hydrogen, 4%. The high interaction rate is characterized by small frequency changes. Even for higher concentrations of hydrogen (3.5–4%), response times are in the range of several seconds (4–8 s), which is very important, especially for the detection of a highly explosive gas such as hydrogen.

The changes in the resistance of the electric module structure (Fig. 5.5) as a result of the hydrogen metering are negligible. The resulting reductions in the differential frequency should be interpreted in the category of mass-elastic interactions. The normal configuration of the frequency modes (see Fig. 4.5a) indicates that the observed effect is the result of an increase in the propagation rate of the SAW, which in turn is the result of reversible changes in the elastic modules of the bilayer structure. The increased mass of the structure, which is the result of the absorption of light hydrogen atoms, is not decisive in this case.

It should be noted that the high speed of operation of the acoustic module is characteristic of reversible changes in the internal structure of the sensor elements used. The resistive method of detecting hydrogen in air (for concentrations 2–10%), using palladium mesoscope structures (*palladium mesowires*), is sometimes characterized by responses below 75 ms at room temperature. Such high speed, observed by decreasing (for hydrogen) and increasing (for air) electrical resistance, results from

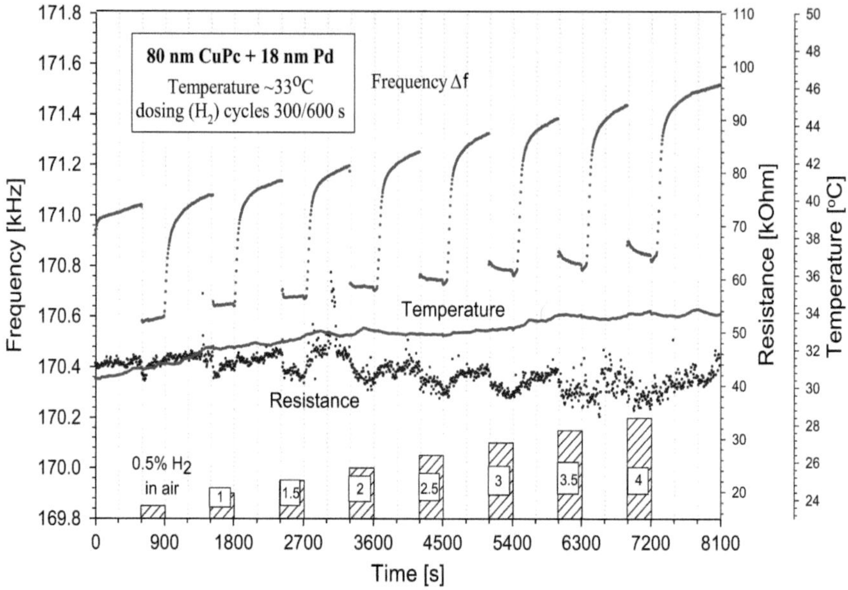

Fig. 5.5 Responses of bilayer structure CuPc 80 nm with Pd 18 nm at ~33 °C for several hydrogen concentrations in air [9]

expansion and contraction of the internal structure of palladium nanowires matrix [3, 27].

The SEM image of the surface of the bilayer structure (100 nm CuPc + 18 nm Pd) made on glass is shown in Fig. 5.6, while its responses with several concentrations of hydrogen in the air are shown in Fig. 5.7.

Fig. 5.6 SEM picture of bilayer structure 100 nm CuPc covered by 18 nm film of Pd on glass substrate

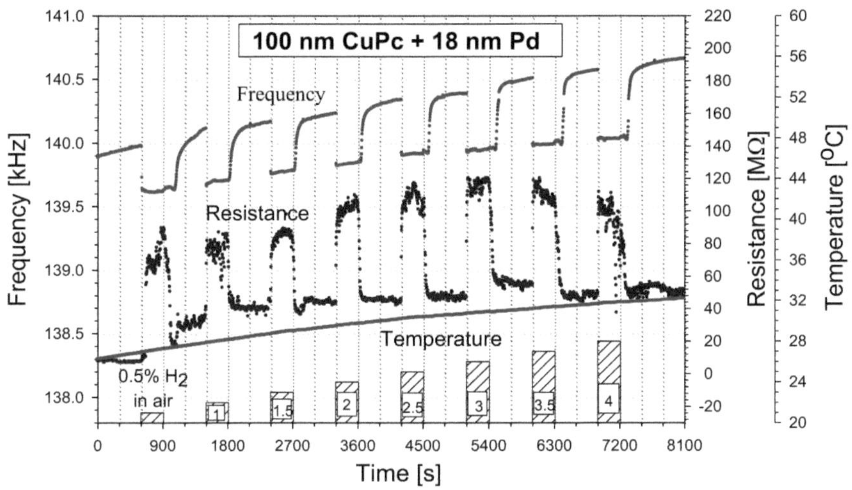

Fig. 5.7 Responses of the bilayer structure CuPc 100 nm and Pd 18 nm at about 28–32 °C with several hydrogen mixtures (0.5–4%) in synthetic air [7]

For a bilayer structure in which the thickness of CuPc ~100 nm phthalocyanin was increased, the change in the differential frequency increased slightly to 600 Hz, for a concentration of 4% and a temperature of 28–32 °C (Fig. 5.7). The structure is characterized by a high rate of interaction with hydrogen for each concentration (less than 8 s—the interval between each measurement). The long-term, slow increase in frequency is associated with a positive temperature drift inside the measuring chamber. Apparent increases in sample resistance, i.e. decrease in electrical conductivity, and coincide with changes in frequency. Analysis of the frequency mode configuration (NMC, i.e. $f < f_0$) indicates that, (see Fig. 5.5), the decreases Δf are caused by increasing the measurement frequency f. In the case of NMC, an increase in the frequency f is possible if the elastic modulus of the structure is increased and/or its electrical conductivity is reduced (see Fig. 4.7). The observed increases in the resistance of the bilayer structure (with AE parameter $\xi = 5.5 \times 10^{-2}$) as a result of interaction with hydrogen may be the cause of acoustoelectric interactions. Small changes in frequency (400–600 Hz) and characteristic high interaction rate indicate that mass-elastic interactions also play a large role.

Figure 5.8 shows the changes in the differential frequency for the bilayer structure at a temperature of ~38 °C and at eight different concentrations of hydrogen in the air. For each interaction, depending on the concentration of hydrogen, there was a repeatable decrease in the frequency of Δf. The structure reacts very quickly to the given hydrogen concentration (4–8 s)—the time between individual measurements. For hydrogen concentrations above 2%, there is a significant increase in the electrical resistance of the structure, caused by the phase transition effect in PdH_x. Acoustoelectric interactions are possible because the decrease in Δf, for the established normal frequency mode configuration (NMC), is the result of, among other things,

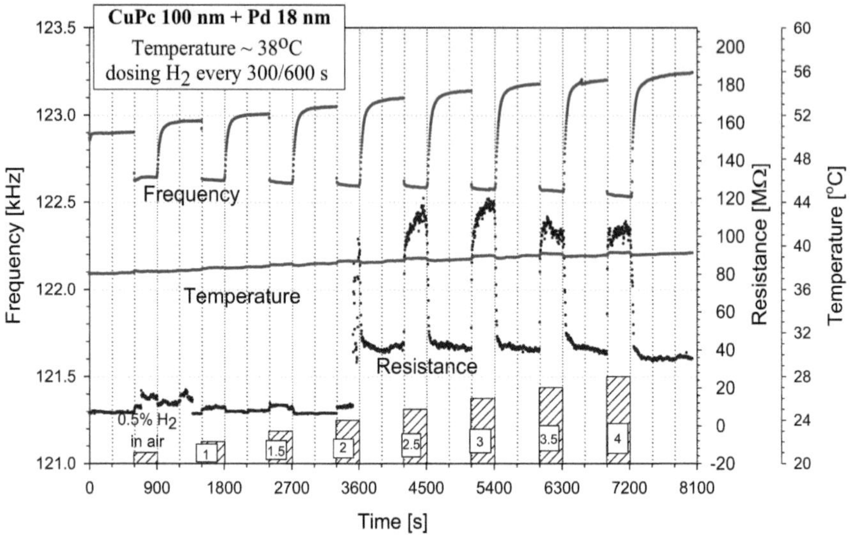

Fig. 5.8 Frequency and resistance changes for bilayer structure CuPc 100 nm + Pd 18 nm at the temperature ~38 °C with hydrogen 0.5–4% in air [9]

a decrease in the electrical conductivity of the structure, i.e. increasing its resistance. The lack of correlation between changes in frequency and resistance for concentrations less than 2% indicates that acoustoelectric interactions are not decisive in this case [9].

On the basis of the acoustic response (Fig. 5.8), a graph of changes in the frequency of the structure as a function of the concentration of hydrogen in the air at a given reaction temperature was made. An almost linear ($r^2 = 0.99$) nature of the changes in the output signal from the concentration of hydrogen was obtained, with the inclination of the characteristic being about 105 Hz/% (Fig. 5.9). The sensitivity of the sensor structure in the studied concentration range, defined as the change in frequency per unit change of concentration, is constant.

5.2.1.2 The Structures of Metal Free phthalocyanine-palladium (H_2Pc-Pd)

Bilayer structures composed of metal-free phthalocyanine (without metal in its macromolecule) and thin layers of palladium were obtained by vacuum evaporation. Several such structures with H_2Pc thicknesses of 80–200 nm and the same palladium thickness of ~20 nm were selected for the most important results. The characteristic features of these sensor structures, compared to CuPc-Pd, are large frequency changes (~kHz) and longer response times under similar measurement conditions, as well as exothermic effects of the reaction of hydrogen with palladium and phase transition effects in PdH_x [23].

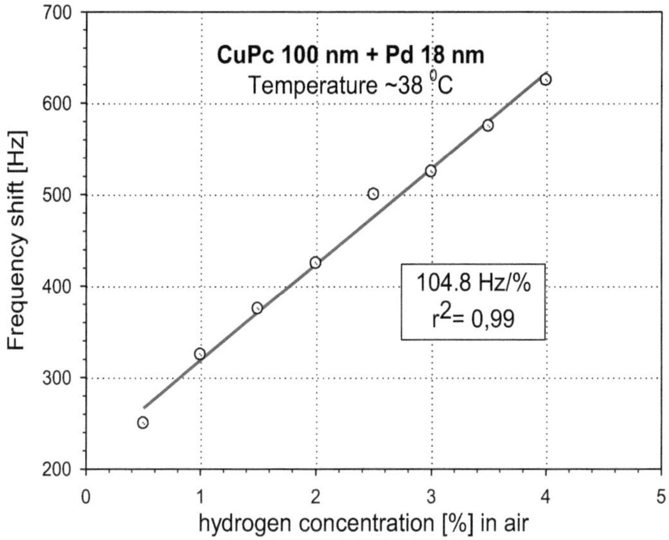

Fig. 5.9 Frequency shifts of CuPc 100 nm and 18 nm Pd bilayer structure at ~38 °C in a function of hydrogen concentration 0.5–4% in air [9]

AFM images of the bilayer structure with hydrogen phthalocyanin (H_2Pc) ~80 nm and palladium (Pd) ~20 nm are presented in Fig. 5.10. The bilayer structures of two sensor modules (acoustic and electrical), manufactured on two different substrates (lithium niobate and glass), in the same technological processes, are currently not as close to each other as for the CuPc-Pd structures. Metal-free phthalocyanine layers are deposited differently on the $LiNbO_3$ substrate and on the glass, because the addition of metal in the phthalocyanine molecule affects its adhesion properties. This is important for the acoustoelectrical method, as the correlation of the results of the two structures cannot be strictly observed.

For the tested bilayer structure, no significant changes in electrical module resistance were found as a result of hydrogen metering in the air at different interaction temperatures. The observed changes in frequency at two temperatures of ~31 and ~48 °C, for two different frequency mode configurations (reverse directions of Δf signal shifts), are not the result of acoustoelectric interactions (Fig. 5.11).

The large frequency changes (~1–2 kHz) are accompanied by local increases in the surface temperature of the sensor structure ΔT ~ +3 °C, visible for a higher interaction temperature of ~48 °C. This effect is associated with the exothermic reaction of hydrogen with palladium and was observed during the first tests of SAW single-layer Pd sensors by D'Amico et al. [2]. The large frequency changes for the tested H_2Pc-Pd bilayer structure are the result of the mass-elastic and thermal effects of the hydrogen-palladium reaction.

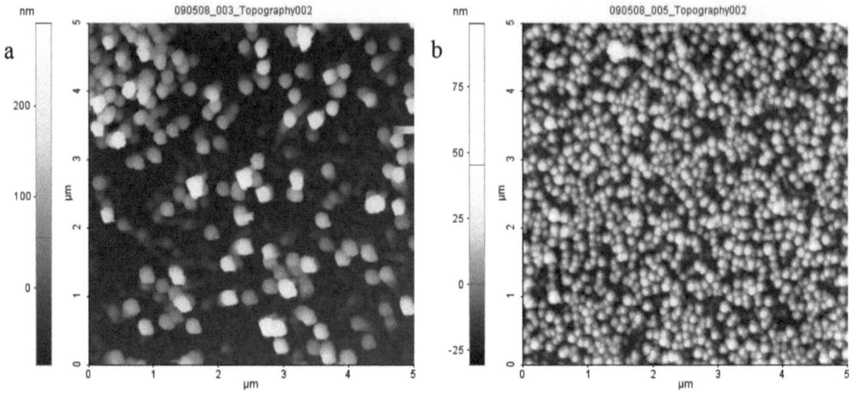

Fig. 5.10 Comparison of AFM 5 × 5 μm pictures of the bilayer structures 80 nm H_2Pc + 20 nm Pd prepared on glass (**a**) and $LiNbO_3$ (**b**) substrates

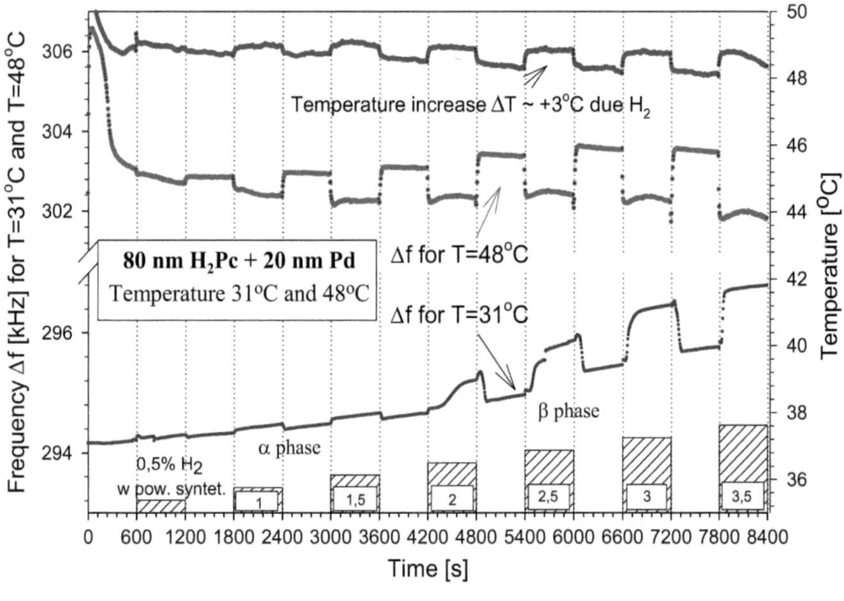

Fig. 5.11 Frequency shifts (at ~31 °C) and frequency shifts and temperature increase (at ~48 °C) for the bilayer structure 80 nm H_2Pc + 20 nm Pd and several mixtures of hydrogen in air

Figure 5.11 also shows the phenomenon of structural changes of the PdH_x compound as a function of hydrogen concentration at a lower reaction temperature of ~31 °C. For low concentrations of hydrogen (<2%), the interactions are very small, which is characteristic for the α-PdH_x phase, whereas, for concentrations above 2%, much larger interactions are visible. The PdH_x is then in the β phase. The concentration of hydrogen at which the phase transition effect occurs depends on the reaction temperature. At higher temperatures, it shifts towards higher concentrations [20] (see Sect. 3.4.3). Similar effects can be observed for the tested structure. At a lower temperature of ~31 °C, the frequency changes are only significant for concentrations above 2% of hydrogen in the air. By contrast, for a higher temperature of ~48 °C, the changes are almost proportional to the hydrogen concentrations. As for the CuPc 720 nm–Pd 20 nm layer structure [11], the phase transition effect shifts with increasing interaction temperature. This is a favorable phenomenon from the point of view of using bilayer structures in practice. The operating temperature of such a structure must be precisely selected in order to obtain proportional frequency changes relative to a certain range of hydrogen concentrations.

Bilayer structures with metal-free (hydrogen) phthalocyanine (H_2Pc) up to 120 nm and palladium (Pd) at 20 nm are characterized by even greater interactions than previous structures. Their interaction with hydrogen at a temperature of ~50 °C is shown in Fig. 5.12. Repetitive changes in the structure's resistance are visible, corresponding to very large frequency changes (~15 kHz for the highest concentrations). The resistance decreases from 25 to 15 MΩ when the structure interacts with hydrogen. Also significant are visible increases in the surface temperature of the structure (+2 °C) for higher hydrogen concentrations, which are the result of the already mentioned exothermic reactions of palladium with hydrogen.

The analysis of the interactions indicates that acoustoelectric effects may occur in the case of an inverse frequency mode configuration (IMC) that occurs here ($f > f_0$). According to the diagram (see Fig. 4.8), the decrease in the differential frequency is

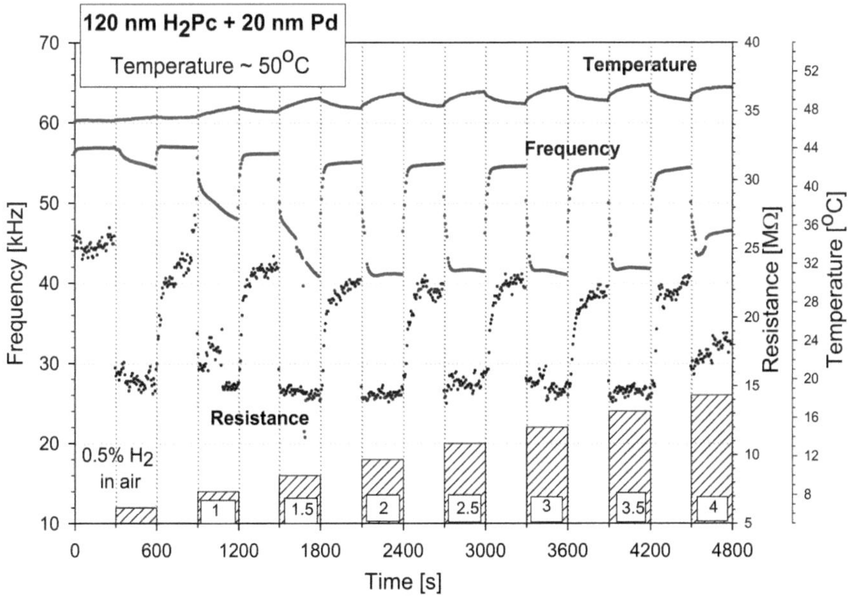

Fig. 5.12 Bilayer structure 120 nm H_2Pc + 20 nm Pd: interaction with hydrogen at temperature increased to about 50 °C [12]

a consequence of a decrease in the propagation velocity of the SAW, which in turn may be caused by an increase in the surface density of the structure, a decrease in their elastic modulus and an increase in the incident surface electrical conductivity σ_S. The observed decreases in the resistance of the structure correspond precisely to the increases in σ_S. A negative frequency temperature coefficient for $LiNbO_3$ Y–Z (~–4 ppm/°C) indicates that the observed very large changes in f may be due to, inter alia, exothermic reactions of hydrogen with palladium. However, the quantitative analysis leads to the conclusion that such large frequency changes cannot be explained by thermal effects. At changes of ΔT ~+2 °C, the corresponding frequency reduction is ~8.2 kHz relative to the observed −15 kHz. There must therefore be acoustoelectric interactions, giving an additional −6.8 kHz frequency shift.

The SEM image of a bilayer structure made on glass with metal-free (hydrogen) phthalocyanine (H_2Pc) with thickness 160 nm and the same palladium thickness (Pd) 20 nm as previously is shown in Fig. 5.13, while its interaction with hydrogen at temperature ~31 °C in Fig. 5.14.

For the studied bilayer structure, the phase transition effect between concentrations of 1.5 and 2.5% of hydrogen in the air at ~31 °C is clearly visible. There is also a very good correlation between frequency changes and resistance changes. Above 1.5%, hydrogen concentration increases in frequency correspond to resistance decreases. The interactions should be interpreted as acoustoelectric. The normal configuration of frequency modes ($f < f_0$) in this case indicates that increases of Δf are caused by decreases in the frequency of the measuring loop (with the sensor structure), which in turn are the result of a decrease in the propagation velocity of SAW.

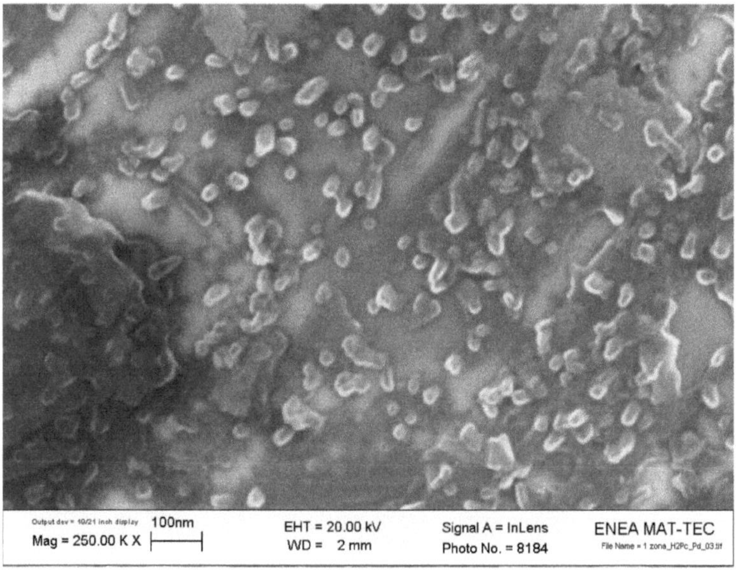

Fig. 5.13 SEM picture of H_2Pc 160 nm + Pd 20 nm structure manufactured on glass

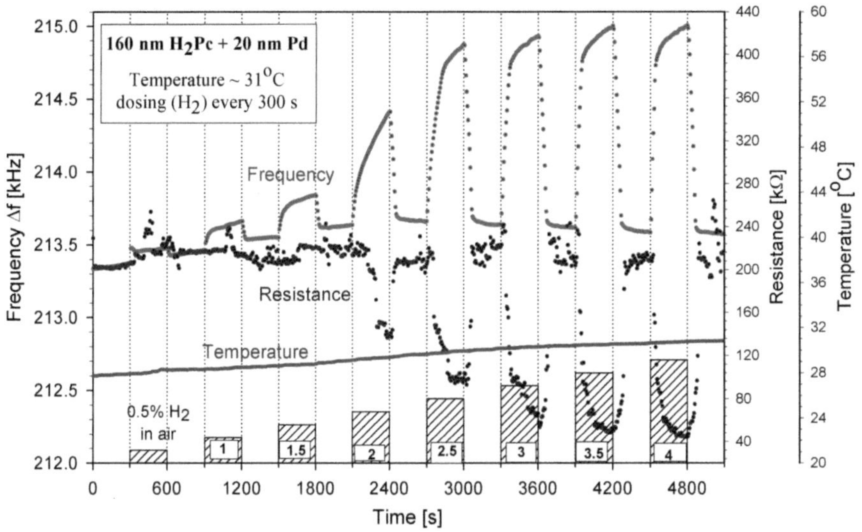

Fig. 5.14 Frequency shifts and resistance decreasing of H_2Pc ~ 160 nm + Pd ~ 20 nm structure at the temperature ~31 °C and several concentrations of hydrogen in air [12]

According to the diagram (see Fig. 4.7), the decrease in SAW speed can be caused by an increase in surface density, a decrease in elastic modulus and/or an increase in the electrical surface conductivity of the structure. The observed reductions in sample resistance, for 4% hydrogen concentration $\Delta R_{sam} = -180$ kΩ, correspond to an increase in the electrical surface conductivity of the structure by approximately $+3.7 \times 10^{-6}\ \Omega^{-1}$. For the resultant acoustic parameter of this structure at an impact temperature of ~31 °C, $\xi_0 = 2.8$, the relative change in SAW velocity can be estimated on the basis of the proposed bilayer structure model (see Fig. 3.17).

In the case of large values of the x parameter (when the conductivity of the second layer is many times greater than the first) for the considered bilayer structure, the read changes ($\delta(\Delta v/v_0)$) from the theoretical curve at an increase of $\xi = 6.4$ (for a concentration of 4% H_2) are about 0.03, which should give a reduction of the measuring frequency of almost 11 kHz. The observed much smaller changes indicate that changes in the elastic modulus of the structure, which in this case reduce the resultant effect, are also a significant cause of the interactions.

The relatively large and repeatable responses led to testing with an additional, prototype polyethylene (PE) (see point 5.3) protective layer for this bilayer structure. These layers should limit the impact of air humidity, for example.

Bilayer structures with hydrogen phthalocyanine (H_2Pc) with the largest thickness of 200 nm and palladium (Pd) 20 nm, interacting with hydrogen at ~33 °C, are shown in Fig. 5.15.

For the bilayer structure with the greatest H_2Pc thickness ~200 nm, the phase transition effect in PdH_x is clearly visible (as in the structure with the greatest CuPc

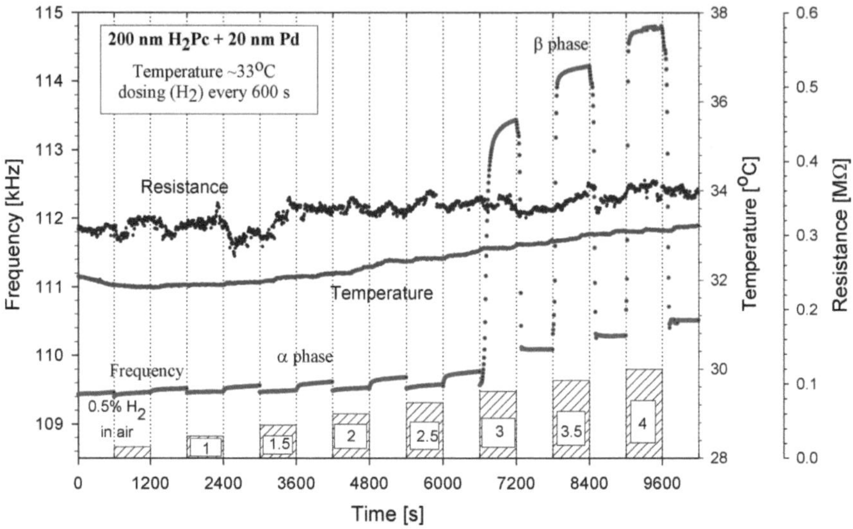

Fig. 5.15 Response of bilayer structure 200 nm H_2Pc + 20 nm Pd at temperature ~33 °C and several hydrogen concentrations (0.5–4%) in air

thickness: 720 nm CuPc–20 nm Pd [11]). In the α phase, for concentrations of 0.5–2.5% H_2, a small (~40–200 Hz) but very rapid shifts in the differential frequency are observed, whereas, in the β phase (>2.5%), its very large (~5 kHz) increase. No changes in the electrical module resistance were observed in this case. The results obtained are not the result of acoustoelectric, but mass-elastic interactions. The inverse mode configuration of the obtained differential frequency of 109.5 kHz, at ~33 °C, is due to the fact that $f > f_0$. As shown in Fig. 4.8, slight decreases in the differential frequency Δf in the range of low hydrogen concentrations (in phase α) are caused by decreases in the measurement frequency f as the velocity of SAW propagation decreases slightly. The decrease in the wave velocity should be interpreted in terms of mass effects, by increasing the surface density of the structure, due to the absorption of hydrogen atoms. However, for hydrogen concentrations above 2.5% for the β phase, the increase in the frequency of Δf results from the fact that the measurement frequency increases as a result of the increase in the speed of the SAW propagation. Large increases in the speed of SAW are caused in this case by a significant disturbance of the elastic modulus of the structure, which is characteristic of the β phase of the PdH_x compound [23].

5.2.2 *Bilayer Structures Metal Oxide-Palladium*

5.2.2.1 NiO_x-Palladium Structure

The NiO_x sensor layer with a thickness of about 60 nm was made by reactive sputtering of nickel target, whose purity was 99.7%, in a controlled atmosphere of Ar:O_2 (3:1). In the magnetron sputtering process, a homogeneous layer with very good adhesion to the substrate was obtained. On the other hand, the Pd layer of ~18 nm was applied in the process of vacuum evaporation from tungsten heater. The application rate was 0.3–0.5 nm/s.

The interactions of these structures with hydrogen are presented in the work of Jakubik et al. [17, 14]. The observed small changes in frequency are typical mass-elastic interactions. The inverse configuration of the frequency modes IMC (f = ~42,965 kHz > f_0 = ~42,939 kHz) results in an increase in the differential frequency as the velocity of SAW propagation increases. The interactions of the NiO_x-Pd structure with hydrogen result from reversible changes in their elastic modulus.

A different type of interaction of the tested structure was recorded for carbon monoxide in the low concentration range of 50 and 100 ppm in dry (RH ~ 4.9%) and humid (RH ~ 50%) air at ~60 °C (Fig. 5.16).

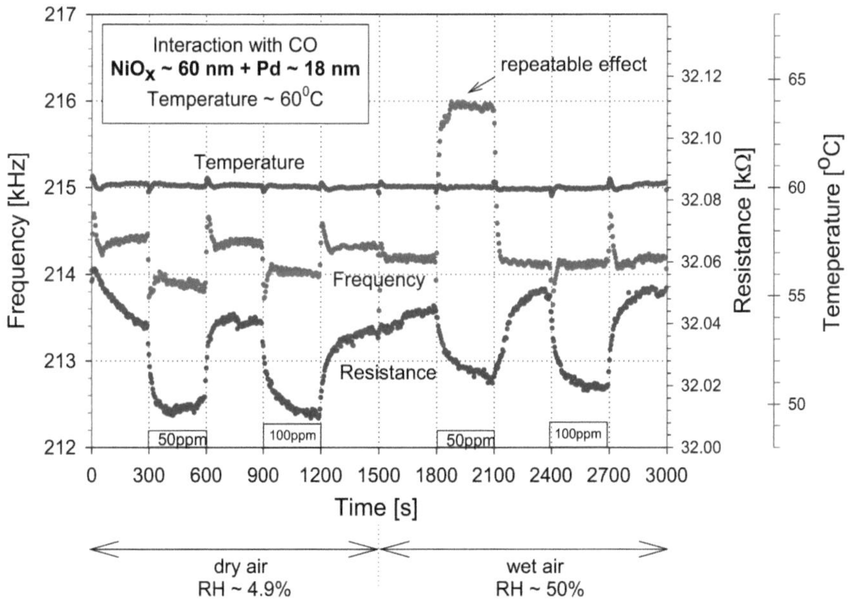

Fig. 5.16 Changes of frequency and resistance for the structure NiO_x-Pd at temperature ~60 °C with carbon monoxide 50 and 100 ppm in sythetic dry (RH ~ 4.9%) and wet air (RH ~ 50%)

In dry air, for the first (50 ppm) and second (100 ppm) CO concentrations, reductions in frequencies of approximately 750 and 500 Hz and resistance of ~0.3 kΩ are observed, respectively. When the atmosphere changes from dry to humid, there is a small decrease in frequency and an increase in the resistance of the structure. For the first CO concentration (50 ppm), a very large increase in frequency of ~1.8 kHz is observed in humid air. This effect is repeatable and occurs in any measurement with similar parameters. On the other hand, dosing CO in humid air causes a decrease in the resistance of structure by 2.5 kΩ and 3.2 kΩ, respectively. Possible acoustoelectric interactions are compensated by changes in the elastic modules. Throughout the measurement process, a strong interaction of the sensor structure with CO is visible, especially in the case of the electrical module.

Very similar interactions as for CO were recorded for ammonia under similar measurement conditions. The interaction of structure with NH_3 at concentrations of 50 and 100 ppm in dry (RH ~ 4.4%) and humid (RH ~ 48%) air at ~60 °C is shown in Fig. 5.17.

Dosing of NH_3 with a concentration of 100 ppm in dry air results in a slight reduction in frequency of about 150 Hz and resistance by 10 kΩ. In the humid atmosphere, on the other hand, the first concentration of NH_3 (50 ppm) causes a strong interaction between the structure and the gas molecules in the form of a large change in frequency of ~1.9 kHz. As for CO, the effect is completely repeatable and occurs in any measurement with similar parameters. On the other hand, for the

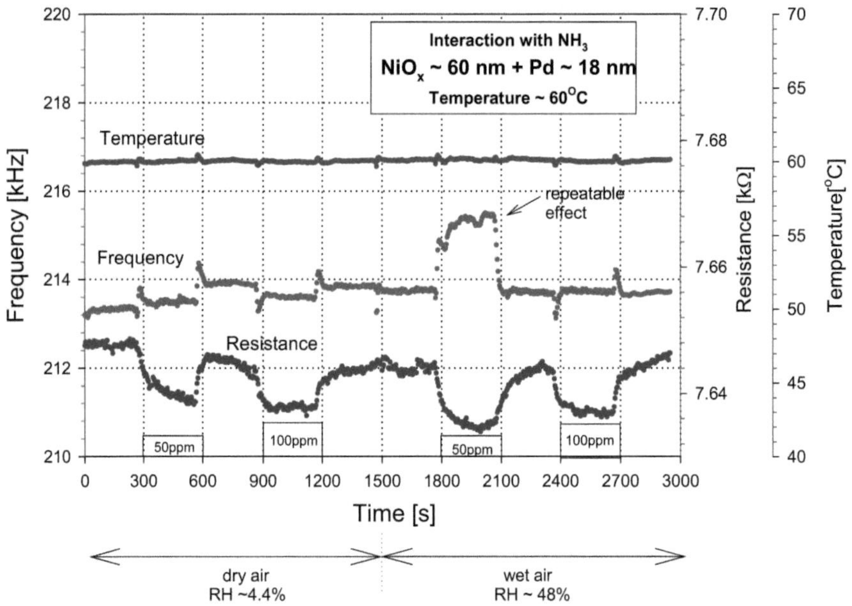

Fig. 5.17 Changes of frequency and resistance of the NiO_x-Pd structure at the temperature ~60 °C with ammonia 50 and 100 ppm in synthetic dry (RH ~ 4.4%) and wet (RH ~ 48%) air

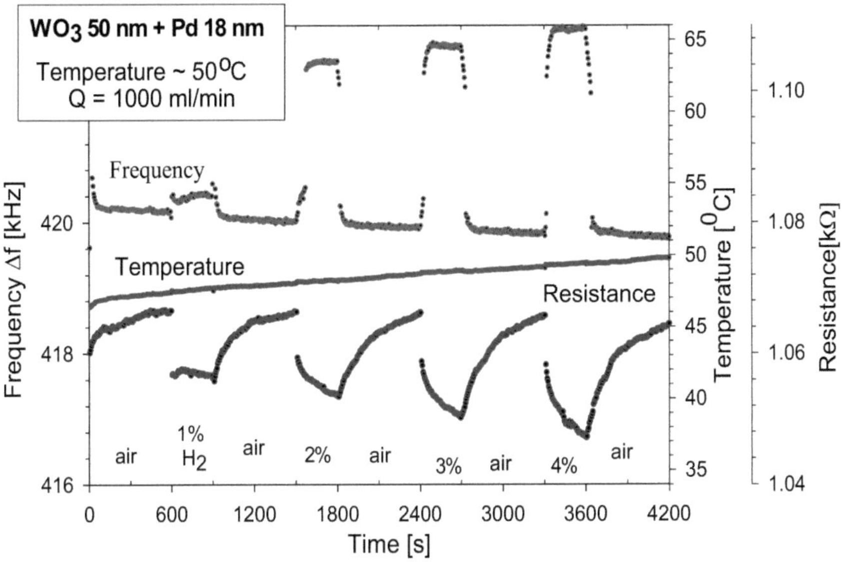

Fig. 5.18 Responses of the 50 nm WO_3–18 nm Pd structure at temperature increased to about 50 °C with hydrogen 1–4% in air (Jakubik 2007b)

second concentration of NH_3, there is no change in oscillation frequency. Possible acoustoelectric interactions as a result of changes in resistance of the structure are compensated by mass-elastic interactions.

5.2.2.2 WO_3-Palladium Structure

For WO_3-Pd structures, fast and large frequency changes were obtained in the hydrogen dosing of 1–4% in air, but at elevated temperatures up to about 50 °C. A very good correlation between acoustic and electrical interactions (reduction of structure resistance) indicates strong acoustoelectric interactions (Fig. 5.18). Obtaining distinct saturation states, which are additionally achieved in a fairly short time (~10–20 s for the α PdH_x phase with hydrogen concentration of 1%, and ~ 100 s for the β phase with concentrations of >1%, made it possible to perform the sensitivity curve of the structure (Fig. 5.19). There is an area of partial saturation above hydrogen concentration of 2.5%, which is associated with the phase transition effect in PdH_x.

5.2.3 The Use of Bilayer Structures in Hydrogen Detection

A summary of the possibilities of using selected phthalocyanine-palladium bilayer structures (MPc + Pd) and metal oxide-palladium (MO + Pd) in hydrogen detection

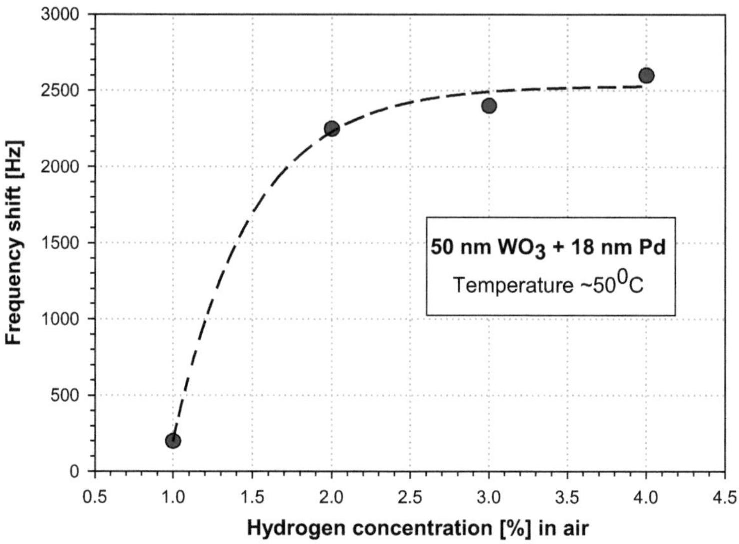

Fig. 5.19 Sensitivity curve for the bilayer structure 50 nm WO_3–18 nm Pd at the temperature increased to ~50 °C with respect to hydrogen in air

Table 5.1 Properties of bilayer structures with SAW for hydrogen detection

Bilayer structure	Maximal frequency changes, Hz	Interaction temperature, °C	Characteristics/ sensitivity	Comments
NiO_x + Pd 60 nm + 18 nm	600 for 2–2.5% H_2	~35	Very nonlinear	Mass interactions
CuPc + Pd 100 nm + 18 nm	650 for 0.5–4% H_2	~38	~Linear/105 Hz/ %	Fast response
H_2Pc + Pd 160 nm+ 20 nm	3800 for 0.5–4% H_2	~40	Nonlinear	Big changes
WO_3 + Pd 50 nm + 18 nm	2600 at 50 °C for 1–4% H_2	~50	Nonlinear	Shifts ~10 kHz! at T ~ 100 °C

is presented in Table 5.1. Hydrogen detection using bilayer structures is possible even at relatively low temperatures.

5.2.4 *Bilayer Structure Nafion®-Polyaniline*

Another type of bilayer structure is the Nafion-PANI (polyaniline + nafion), which was investigated using the acoustoelectrical method. Due to the very high resistance of the structure, a Keithley electrometer was used in the research. The main purpose of the Nafion application was possible protonation of the second polyaniline layer to increase its electrical conductivity.

Nafion® with a thickness of ~300 nm was applied by a spin-coating method at a speed of ~7500 rpm. The layer was heated at 40 °C for 2 h and then for 15 min at 120 °C to harden it. A layer of polyaniline with a thickness of 72 nm was produced by a vacuum evaporation process. The whole process was carried out in a residual Ar atmosphere, due to the oxidation of the polymer during its application, which causes deterioration of the sensor properties. To obtain the desired atmosphere, the vacuum chamber had to be rinsed three times with argon. Then, to evaporate the water trapped in the PANI, the steam source was heated at 200 °C for 10 min. The polyaniline deposition process was carried out at 350–400 °C. The deposition (sublimation) of the desired thickness took ~45 min. The interaction of Nafion-PANI structures with NH_3 concentrations of 50 and 100 ppm in dry (RH ~ 4.9%) and humid (RH ~ 48%) air at ~31 °C is shown in Fig. 5.20.

In dry air (0–1500 s), the structures almost do not interact with ammonia. No changes in frequency were recorded, while the resistance decreases very slowly in the range of 0.2 GΩ for a higher concentration of 100 ppm NH_3. When the surrounding atmosphere changes from dry to humid, there is a decrease in frequency of approximately 500 Hz and a very large decrease in resistance of approximately 0.85 GΩ. For the first (50 ppm) and the second (100 ppm) NH_3 concentrations in humid air, there is an increase in frequency of approximately 300 Hz and 400 Hz, respectively, and a decrease in resistance of ~50 MΩ. The normal configuration of

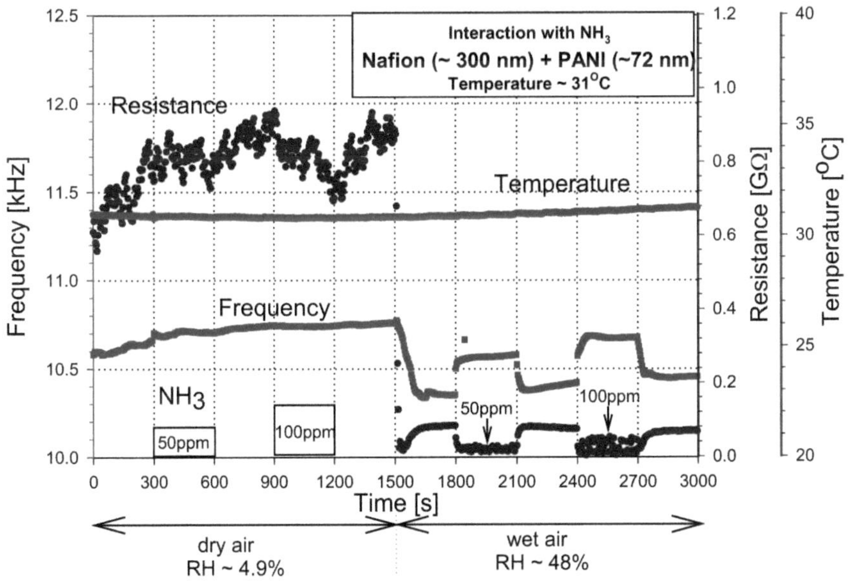

Fig. 5.20 Responses of the bilayer structures Nafion®-PANI with NH_3 at the temperature ~31 °C in synthetic dry air (RH ~ 4.9%) and wet air (RH ~ 48%) [8]

the frequency modes (NMC $f < f_0$) indicates (see Fig. 4.7) that the observed changes in frequency may be the result of an increase in the surface density of the structure (absorption of NH_3 molecules), as well as diminishing of elastic modules and an additional acoustoelectric effect. It can be concluded that the presence of water vapor significantly influences the formation of significant interactions (resistance changes) between the bilayer structures of nafion-polyaniline and ammonia. Such phenomena have also been observed by other authors [21].

No significant interactions were found for this type of bilayer structure with hydrogen and carbon monoxide, whereas the influence of low concentrations of nitrogen dioxide at ~32 °C was quite significant, as for ammonia (Fig. 5.21). As for ammonia, in the dry atmosphere there is a complete absence of interactions for both lower (200 ppm) and higher (400 ppm) NO_2 concentrations. Similarly, with the change in atmosphere from dry to humid, there is a decrease in frequency of about 400 Hz and 0.8 GΩ. In humid air, there is an increase in frequency of about 200 Hz and 250 Hz, respectively, for each concentration (200 and 400 ppm). For the first dose of NO_2, the resistance decreases similarly as for ammonia by ~50 MΩ. The existing mass-elastic interactions (for NMC of the frequency) are similar as for ammonia. Some relaxation phenomena can be observed for the second concentration of 400 ppm NO_2 in a humid atmosphere for the acoustic response. The saturation state is not reached because the frequency signal tends to decrease systematically. This is characteristic of interactions that compensate for each other.

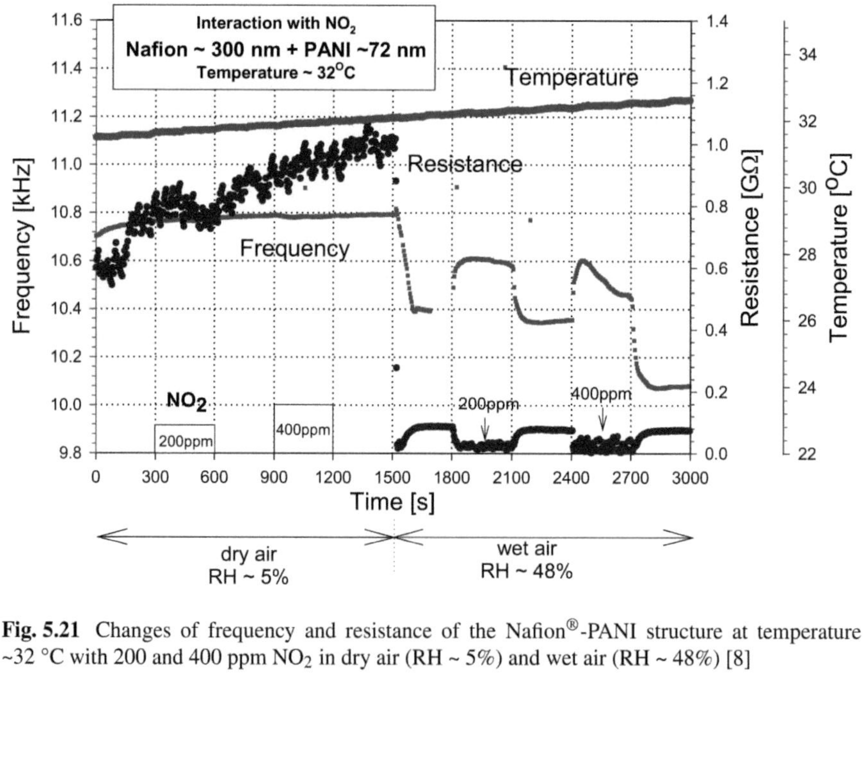

Fig. 5.21 Changes of frequency and resistance of the Nafion®-PANI structure at temperature ~32 °C with 200 and 400 ppm NO_2 in dry air (RH ~ 5%) and wet air (RH ~ 48%) [8]

Acoustoelectric interactions, in the case of decreasing the resistance of the structure (increasing its surface electrical conductivity), and frequency NMC, can cause an increase in the differential frequency, visible for NO_2 dosing. On the other hand, the observed 400 Hz decrease Δf, as a result of increasing humidity of the surrounding atmosphere, cannot be the result of AE interactions alone. In the case of NMC and reducing the resistance of the structure, the Δf frequency would increase. The increase in air humidity must have caused a change in the elastic modulus of the structure, due to the presence of water molecules in the surrounding atmosphere.

5.3 Multilayered Structures

Making a properly selective sensor structure is one of the most important issues in the design of gas sensors. Such a structure should interact strongly with the molecules of the gas for which it is intended, but weakly, and preferably at all, with the molecules of all other gases, including in particular water vapor. The presence of water vapor is a strong interfering factor in gas sensors with SAW. Although the effect of condensation of water molecules on the transducers of the measuring system is compensated in the differential system, it cannot be avoided in the case of a reaction with the sensor structure. One solution to increase the selectivity of sensor structures to some extent is to make various types of additional molecular membranes that act as protective layers. In this way, sensor structures become multilayer systems.

5.3.1 Additional Polyethylene Layers

A significant reduction in the effect of water vapor on the sensor response with SAW was achieved by applying additional polymer layers of polyethylene (PE) for H_2Pc-Pd bilayer structures. The polyethylene spray occurred at a temperature of about 190–250 °C, and the thicknesses achieved were of the order of several tens of nanometers (up to ~60 nm). Polyethylene layers with different thicknesses: ~30 and ~60 nm were designed to increase the reduction of water vapor impact on the 160 nm H_2Pc–20 nm Pd structure. A thicker polyethylene layer reduces the effect of ~47% of relative air humidity by about 18 times more compared to a thin layer, while the interaction of 2% hydrogen concentration is only slightly less (Fig. 5.22).

Similarly, the influence of low carbon monoxide concentrations was eliminated completely [13]. Without PE layers, the maximum frequency change was 400 Hz for 20 ppm CO, whereas after the formation of a thicker PE layer (~60 nm) there was no frequency change at all, even for higher CO concentrations and higher impact temperature. The influence of ammonia could not be eliminated completely, but it was significantly limited at similar impact temperatures (Fig. 5.23a and b). The maximum frequency changes of ~200 Hz were reduced to about 20 Hz for the same concentration of 500 ppm NH_3.

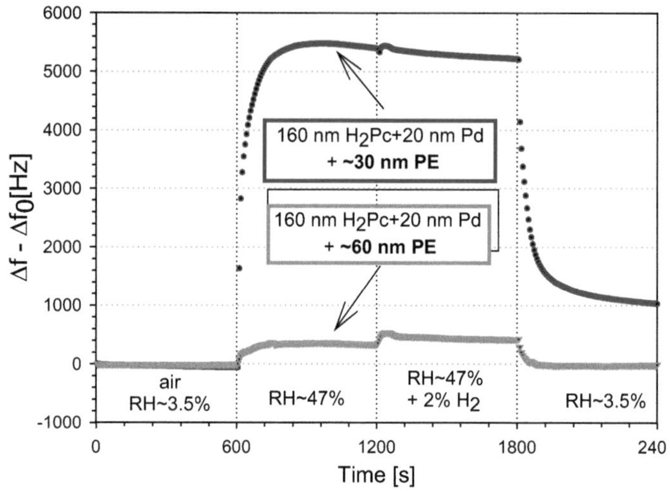

Fig. 5.22 Frequency difference of Δf signal (for wet air ~47% and hydrogen 2%) and Δf_0 (for dry air) for the bilayer structure 160 nm H_2Pc–20 nm Pd covered by polyethylene films (PE) [13]

The detection of hydrogen in such a protected layer structure is still possible, even in an atmosphere of high humidity (Fig. 5.24). However, the frequency changes for the concentration of 4% H_2 are much smaller (~150 Hz) than for unprotected structures (~1200 Hz) (see Fig. 5.14).

5.3.2 Additional TPX Layer

The completely new protective layer is made of a commercially available thermoplastic polymer with the proprietary abbreviation TPX (multimethylopentene) (Mitsui Chemicals). This is a polymer with unique physicochemical properties due to the attached side chains. Its basic characteristics are: low density of 0.83 g/cm^3, good dielectric properties, excellent moisture resistance, and poor permeability to gases. For these reasons, it was decided to use this type of polymer material as an additional protective layer.

A prototype TPX layer was made on a bilayer structure used for the detection of hydrogen 50 nm WO_3–18 nm Pd. After initial (two-fold) purge with argon, at a temperature of 120–160 °C and application time of about 2 min, at an average rate of 0.3 nm/s and a substrate distance of 13 cm from the source, a sensor structure with a TPX protective layer of about 30 nm thickness was obtained. For the bilayer structure with tungsten oxide and palladium, the effect of humidity change of the surrounding atmosphere was much smaller than for structures containing phthalocyanine, even without a protective layer (Fig. 5.26). When the humidity level was changed from

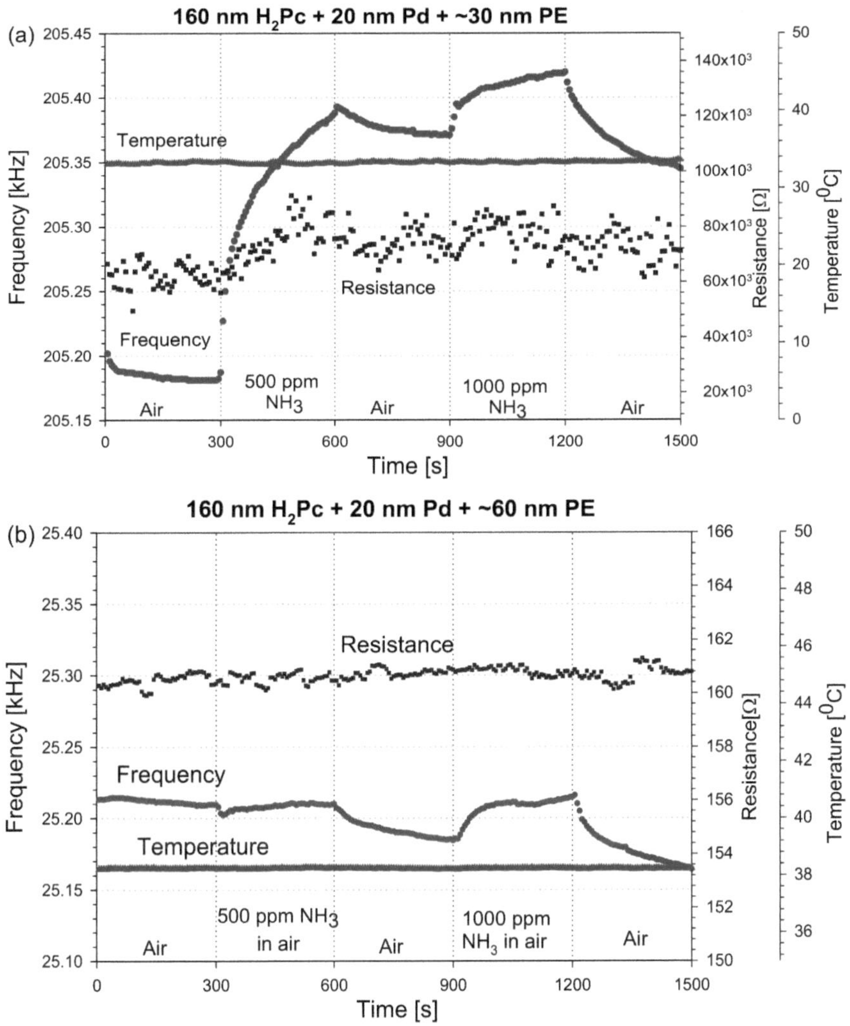

Fig. 5.23 Influence of ammonia concentrations 500 and 1000 ppm in air on multilayered structure 160 nm H_2Pc + 20 nm Pd + ~30 nm PE (**a**), reduction of interaction with NH_3 after preparation of thicker ~60 nm PE film (**b**) [13]

5 to 50%, the frequency increased by ~300 Hz. The use of the ~30 nm TPX layer allowed this effect to be further reduced to ~150 Hz (Fig. 5.27).

The result of the interaction of the WO_3-Pd structure with carbon monoxide without the protective layer is shown in Fig. 5.28, while with the 30 nm TPX layer is shown in Fig. 5.29. The CO effect was significantly reduced, especially for higher

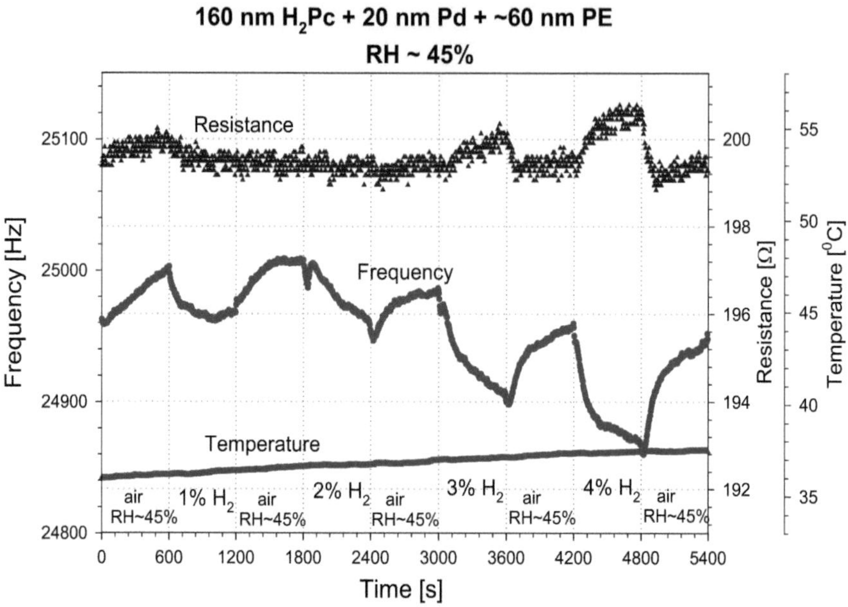

Fig. 5.24 Responses of multilayered structure 160 nm H_2Pc + 20 nm Pd + ~60 nm PE at temperature ~37 °C with hydrogen 1–4% in air with relative humidity ~45% [13]

Fig. 5.25 Structural formula of the multimethylopenthene (TPX)

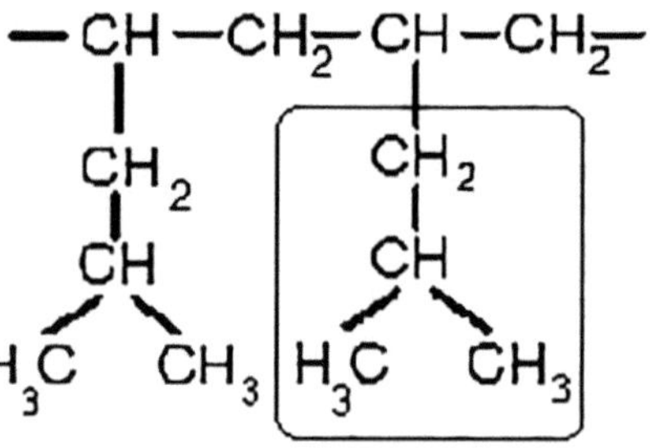

concentrations in the humid atmosphere. It is possible to improve the selectivity of such a layer structure by using additional TPX layers.

The application of an additional TPX protective layer in the WO_3-Pd structure did not reduce its absorption properties relative to hydrogen. Even dosing of relatively low H_2 concentrations (400 and 800 ppm) allowed the observation of frequency changes at 250 Hz for 400 ppm and 800 Hz for 800 ppm, but only when the sample temperature was raised to about 70 °C (Fig. 5.30).

The use of additional protective layers creates opportunities to increase selectivity towards certain gases. For phthalocyanine-palladium bilayer structures, the use of polyethylene layers significantly reduces the impact of water vapor. For WO_3-Pd structures, on the other hand, the use of TPX (multimethylopentane) layers creates

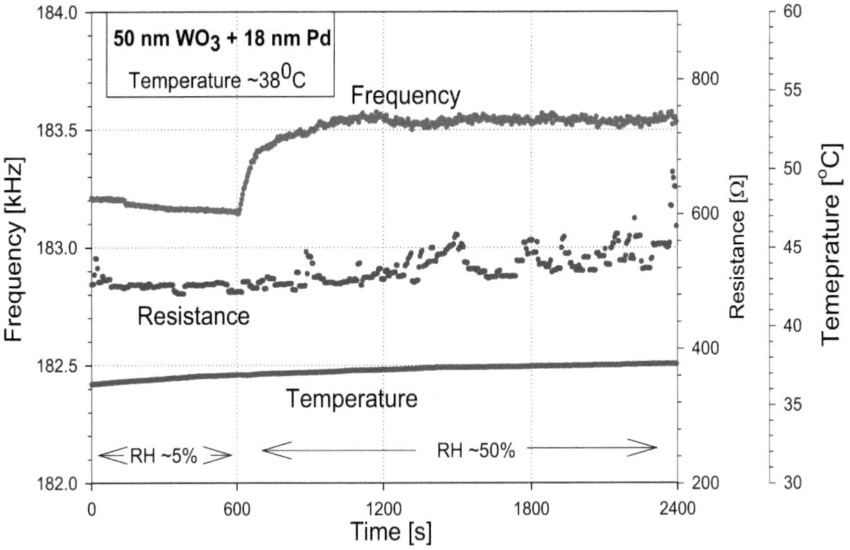

Fig. 5.26 Responses of the bilayer structure 50 nm WO_3 + 18 nm Pd for air RH change 5–50%

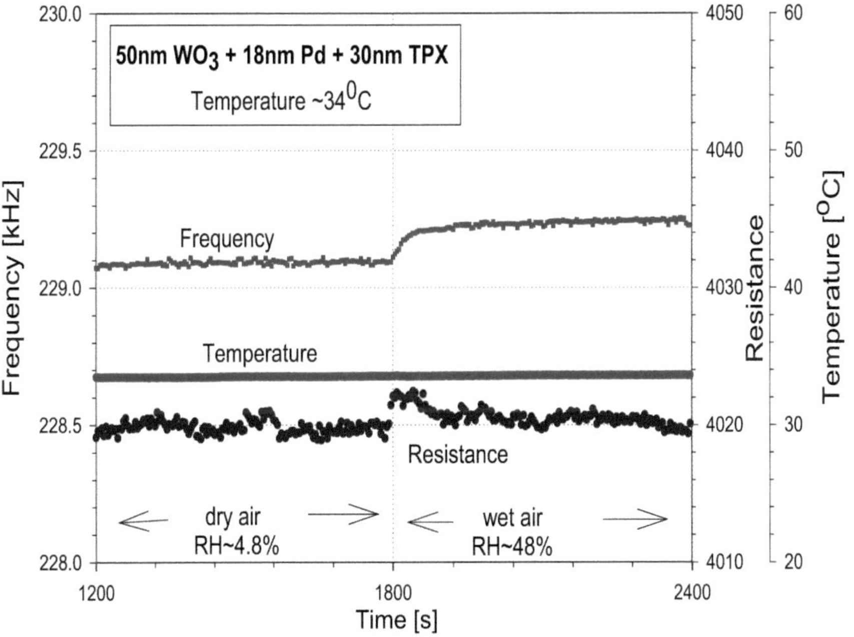

Fig. 5.27 The influence of relative humidity change 4.8–~48% for multilayered structure (50 nm WO_3 + 18 nm Pd + 30 nm TPX)

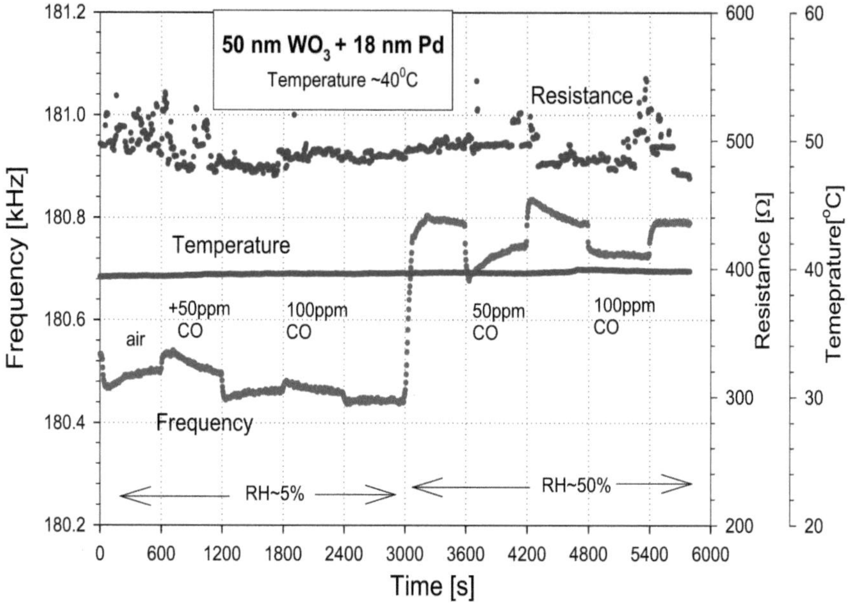

Fig. 5.28 Responses of the bilayer structure 50 nm WO_3 + 18 nm Pd without TPX film with carbon oxide (50 and 100 ppm) in dry and wet air atmosphere

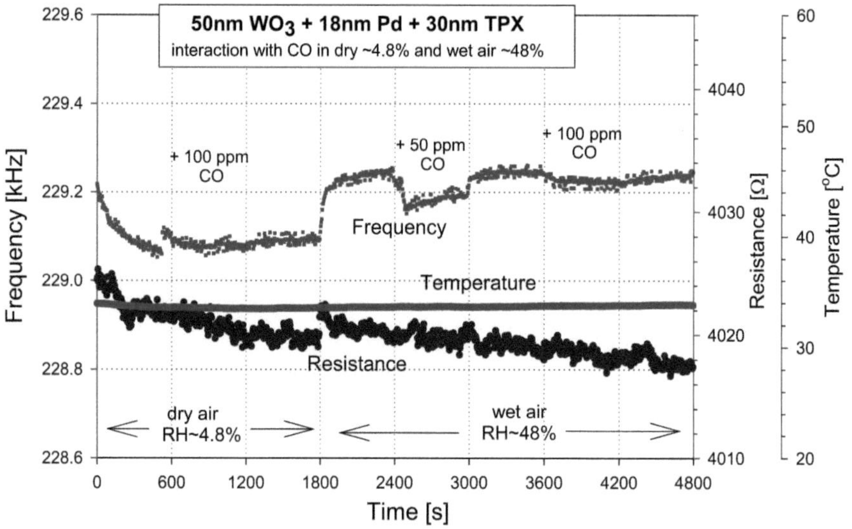

Fig. 5.29 Responses of multilayered structure 50 nm WO_3 + 18 nm Pd + 30 nm TPX with carbon oxide (50 and 100 ppm) in dry and wet air atmosphere

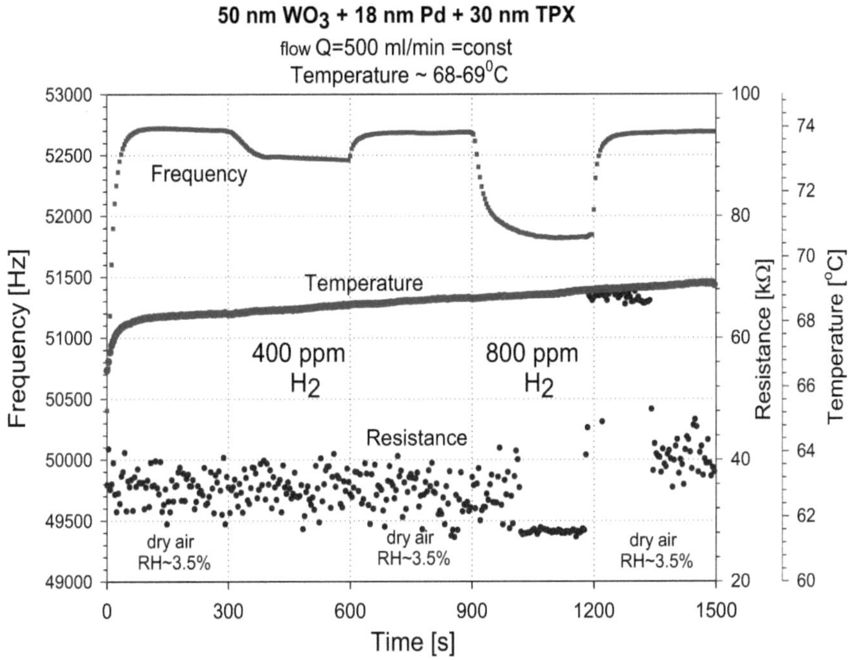

Fig. 5.30 Responses of the multilayered structure 50 nm WO_3 + 18 nm Pd + 30 nm TPX with a low concentrations of hydrogen (400 and 800 ppm) in a dry atmosphere and temperature ~69 °C

opportunities to improve their selectivity without reducing their absorption properties towards hydrogen. However, higher operating temperatures should be used.

5.4 The Results of the Tests Using the Speed of Interaction

In this part of the book, the possibilities of using a new, original detection method, based on the rate of change of the frequency signal in the SAW sensors will be presented. The essence of this method is to use the rate of change in the frequency signal during the interaction of a given sensor structure as a function of different gas concentrations. The genesis of the method stems from the observation that for a certain type of layer structures and higher ranges of hydrogen concentration (up to the explosive limit of 4%), linear areas of change of the frequency signal appear for the bilayer sensor system with SAW. Although the absolute frequency changes for such concentration ranges reach almost the same level [10], the differences in the rate of change of frequency signal for many tested sensor structures at a given hydrogen concentration were very clear. The observed rate differences for the 110 nm H2Pc–20 nm Pd bilayer structure ranged from 2.1 Hz/s for 1.5% hydrogen concentration

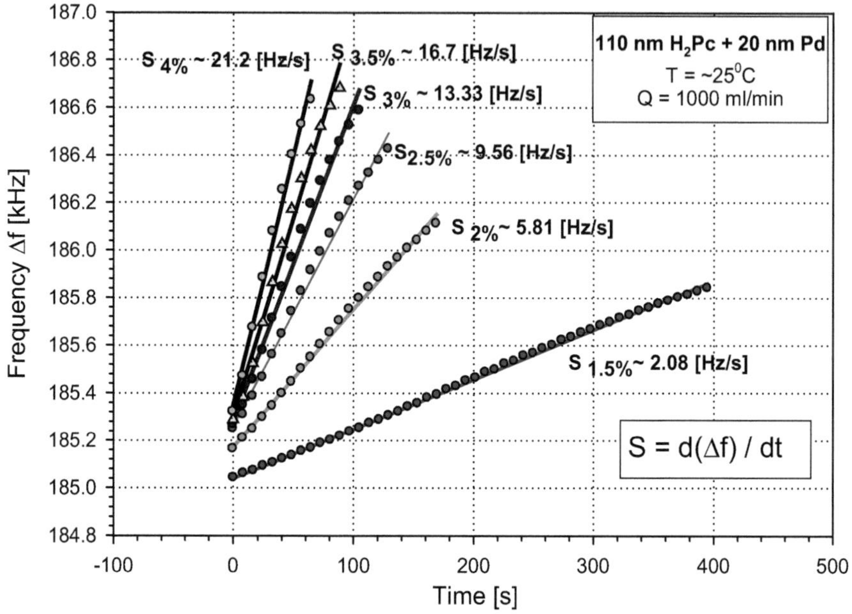

Fig. 5.31 Frequency changes as a function of time at the initial part of the signal increase, for the bilayer structure 110 nm H_2Pc + 20 nm Pd at temperature ~25 °C and several hydrogen concentrations in air—various rates *S* for Δf increase [10]

to 21.2 Hz/s for 4% concentration at ~25 °C (Fig. 5.31). Thus, for the tested range of hydrogen concentrations, the rate of interaction increases tenfold.

The rate of interaction is a linear function of the concentration range of hydrogen concentrations tested, it is higher for interaction in air than in nitrogen. The differences between these rates increase for higher concentrations of hydrogen. The lack of oxygen during interaction of a structure with hydrogen reduces the rate of increase of the frequency signal [10].

The results of the study based on the reaction rate for the bilayer structure with a greater thickness of hydrogen phthalocyanine (160 nm) are presented in Jakubik et al. [12]. On the other hand, the comparison of the reaction rate for two bilayer structures with different thickness of hydrogen phthocyanine (80 and 160 nm) and the same thickness of palladium (~20 nm) at a temperature of ~30 °C, and hydrogen in the air is presented in Fig. 5.32.

In a detection method based on the rate of increase of the frequency signal, there must be linear ranges of changes in this signal relative to different gas concentrations. This method was mainly used for phthalocyanine-palladium bilayer structures, where the most promising results were obtained for structures with hydrogen phthalocyanine covered by a thin palladium layer.

This method is also the subject of a patent [25].

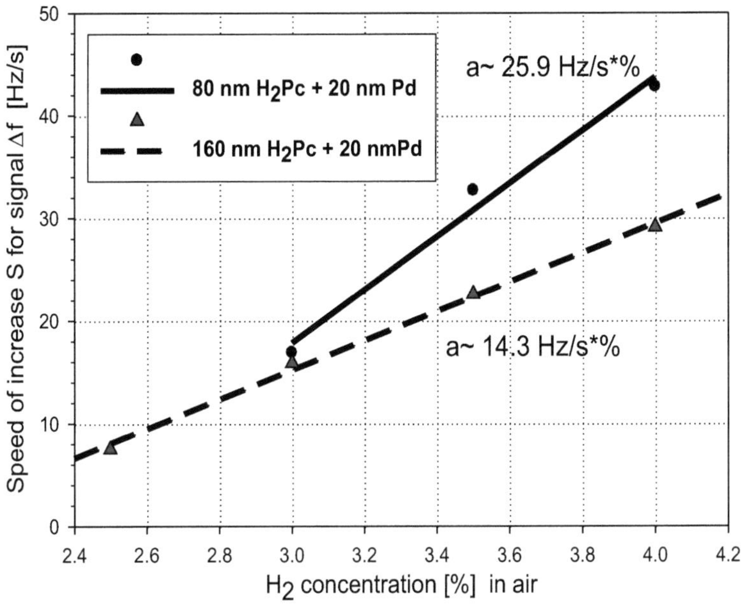

Fig. 5.32 Comparison of interaction speeds for hydrogen in air for two structures 80 nm H_2Pc + 20 nm Pd, and 160 nm H_2Pc + 20 nm Pd at the temperature ~30 °C [12]

5.5 Research Results in Multi-channel System

The quest to create matrix systems resulted in the construction of a multi-channel system on a piezoelectric substrate of the same size as a two-channel system. The four-channel system allows three different sensor structures produced in three measurement tracks to be studied in the same measurement cycle. Track 4 traditionally serves as a reference line, compensating for the undesirable effects of temperature and pressure changes.

Figure 5.33 shows the responses of a prototype four-channel system with three different WO_3-Pd bilayer structures with different thickness of WO_3 (50, 100, and 150 nm) and the same palladium thickness of ~10 nm for decreasing and increasing concentrations of hydrogen (0.5–2%) in the air, and at an increased temperature to ~75 °C. The resulting frequency changes are greater for the structure with the highest thickness of WO_3 and up to 10 kHz for 2% hydrogen (channel 3 Δf_3).

Figure 5.34 shows the effect of the same four-channel module for much lower concentrations of hydrogen in the air (1000–6000 ppm). To make the effects visible, the temperature was raised to 100 °C.

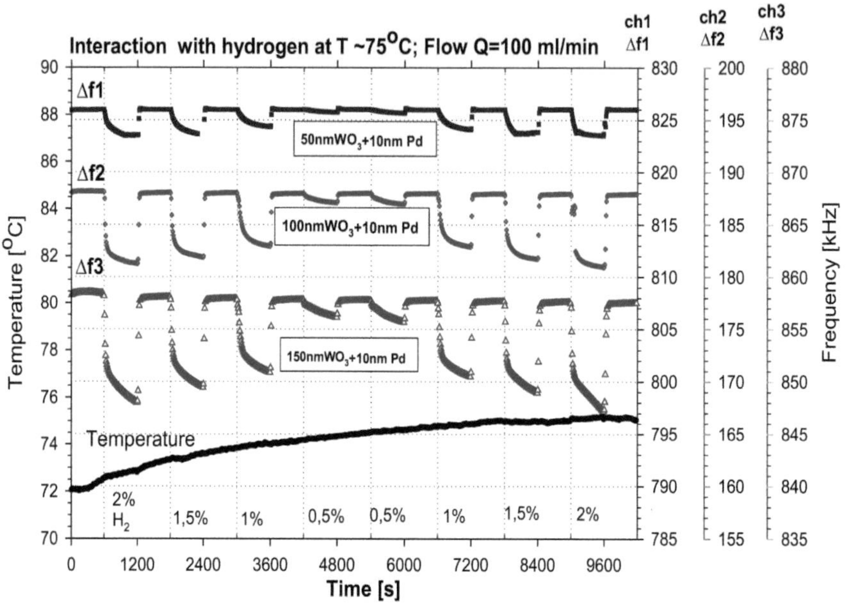

Fig. 5.33 Responses of three bilayer structures of WO_3 (50 (blue curve), 100 (red), and 150 nm (green)) with palladium (Pd) ~10 nm at temperature ~75 °C for decreasing and increasing hydrogen concentrations 0.5–2% in air

Similarly, as before, the largest frequency shifts (~+5 kHz for 6000 ppm H_2) were recorded for the layer structure with the largest thickness of WO_3 (150 nm). However, the saturation state is reached faster for the structure with a smaller thickness of WO_3 (100 nm). The response, however, is then smaller ~+3 kHz for the same concentration of hydrogen (6000 ppm). It is possible to achieve a resolution of hydrogen concentration below 400 ppm, which is visible for increasing concentrations, especially for the structure with the largest density of WO_3. The detection limit of the tested bilayer structure at 100 °C is also of the same order (Fig. 5.35). A measurable shifts in frequency occurs for concentrations above 400 ppm H_2 in air for the structure with the greatest thickness of WO_3 (150 nm).

Figure 5.36 shows the results of tests of bilayer structures of WO_3-Pd type with a mixture of three different gases: hydrogen, methane, and carbon monoxide with different concentrations. The tests were carried out in five groups: I—single gases, II—decreasing hydrogen concentration at constant methane concentration 100 ppm, III—constant hydrogen concentration 0.4% and decreasing methane concentration (50 and 20 ppm), IV—mixture of H_2 and CH_4, and V—mixture of CH_4 and CO.

The tested bilayer structures interact differently with the dosed mixtures of gases. Between groups I (single gases) and III (constant hydrogen concentration of 0.4% and decreasing methane concentration of 50 and 20 ppm) there are differences in response shape. For group V, there is compensation, i.e. no change in frequency. To take advantage of this kind of information, much more research is needed, taking

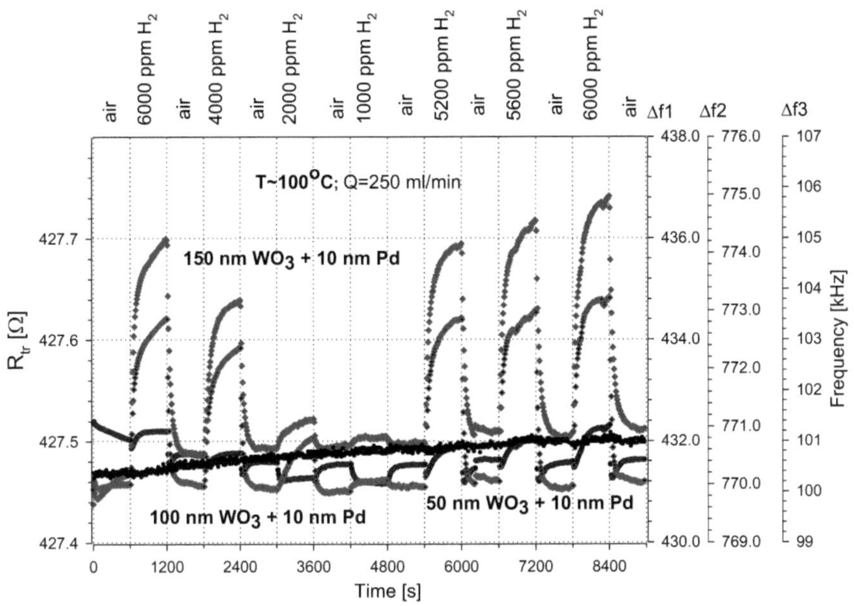

Fig. 5.34 Frequency shifts for three bilayer structures WO_3 (50, 100, 150 nm) + 10 nm Pd at temperature ~100 °C with hydrogen concentrations 6000–1000 and 5200–6000 ppm in air, R_{tr} is a resistance of a thin film thermoresistor performed on the module surface

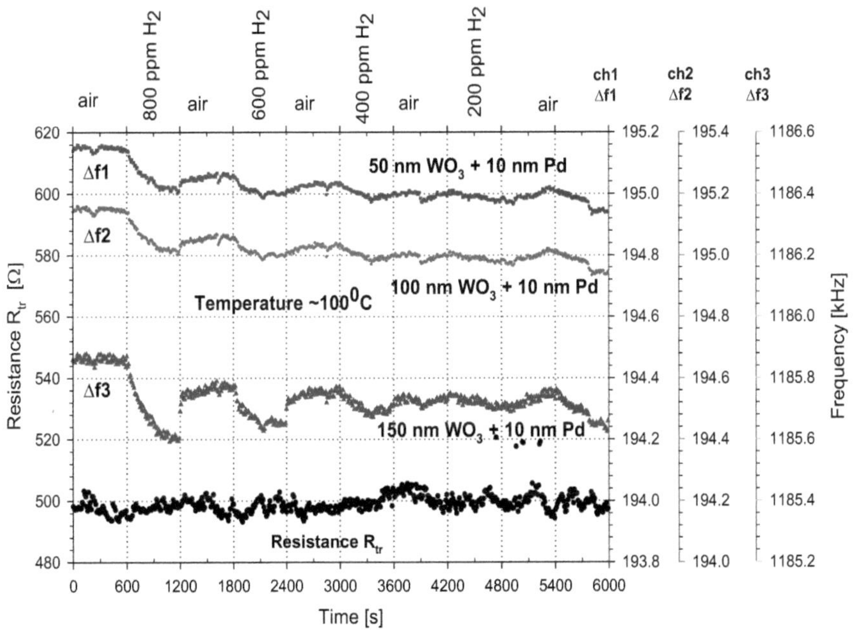

Fig. 5.35 Limit of detection (LOD) for hydrogen and bilayer structures WO_3-Pd at temperature ~100 °C

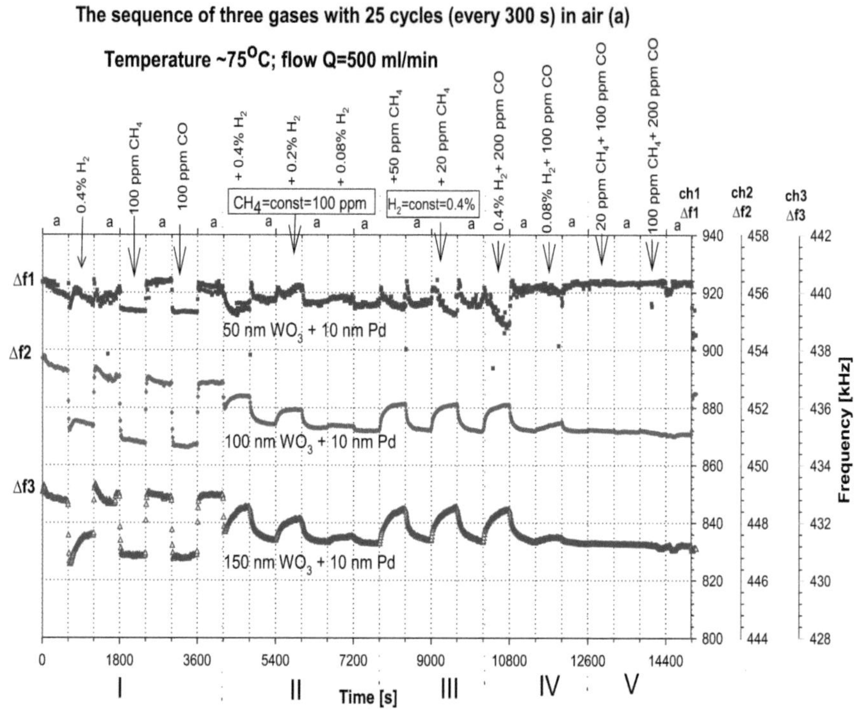

Fig. 5.36 Investigations results for four-channel module with three bilayer structures of WO_3-Pd at temperature ~75 °C, and the sequence of three gases: hydrogen, methane, and carbon oxide

into account different temperatures, humidity, and variable concentration ranges of different gases. This will allow for the effective use of various methods of pattern recognition, e.g. Principal Component Analysis (PCA) and Artificial Neural Network (ANN) methods [24].

5.6 Acoustoelectrical Studies of Bilayer Structures

The main objective of the acoustoelectrical tests was to confirm the simultaneous shifts in frequency (and thus also the speed of SAW) and electrical conductivity of the studied sensor structures as a result of their interaction with gases. The correlation between the results of this type of study indicates the possibility of the occurrence of acoustoelectric interactions, especially for structures made in the same technological processes and tested under the same measurement conditions.

Correlations of this kind have already been shown in some cases for the following bilayer structures: H_2Pc + Pd (see Figs. 5.12 and 5.14) and for WO_3 + Pd (see Fig. 5.18). In addition, resistance tests (in planar electrical system) were performed for several structures before and after the application of palladium. In all cases, there was a decrease in resistance, i.e. an increase in electrical conductivity (even up to six orders of magnitude, e.g. for H_2Pc + Pd), which significantly increases the possibilities of acoustoelectric interactions for semiconductor-palladium bilayer structures.

Table 5.2 presents the obtained test results for the electrical conductivity of single layers and double structures, as well as the acoustoelectric parameter $\xi = \sigma_s / v_o C_s$ (for single layers) or $\xi_1 = \sigma_{s1} / v_o C_s$ (for bilayer structures), which describes the position of a given layer or layered structure on the curves of acoustoelectric interactions. For three practically obtained bilayer structures with design parameter values $x = 1.8$ for WO_3 + Pd, $x = 10^3$ for CoPc + Pd, and $x = 10^6$ for H_2Pc + Pd (where $x \cong \sigma_{s2}/\sigma_{s1}$ is roughly the ratio of electrical conductivity of the layers), the changes in acoustoelectric parameters resulting from the performance of an additional palladium layer on a single layer are shown in Fig. 5.37.

In addition to a single palladium layer (with $\xi \sim 10^5$), for all structures shown in Table 5.2, the AE parameter ξ is less than 1, so the working point is shifted to the left relative to the center of the characteristic. The closest to its center is the WO_3-Pd structure, and for it the most significant effects of the acoustoelectric interactions were obtained. Increasing the temperature of the interaction in most cases shifts the working point towards higher values of the parameter ξ, and thus increases the possibilities of acoustoelectric interactions, which was also observed.

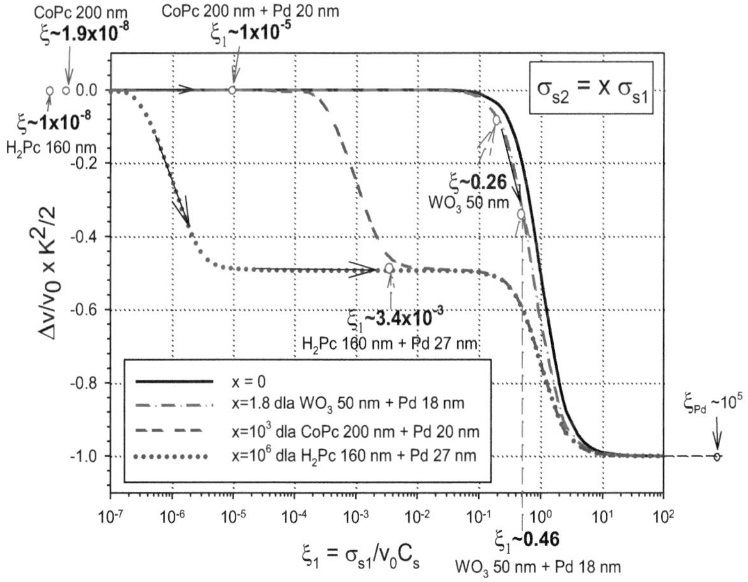

Fig. 5.37 Changes of the AE parameters as a result of the creation of the thin palladium film for three different bilayer structures with a construction parameters $x = 1.8$, 10^3, and 10^6

Table 5.2 Surface electrical conductivity σ_s and parameters x and ξ for the selected single and bilayer structures

Sensor structure	Layer thickness [nm]	Temperature [°C]	σ_s [Ω^{-1}]	Relative increase of σ_s due to Pd film, parameter $x \cong \sigma_{s2}/\sigma_{s1}$	AE parameter ξ or ξ_1
Palladium	20	25	0.185[a]		~10^5
CoPc	200	25	3×10^{-14}		1.9×10^{-8}
CoPc + Pd	200 + 20	25	1×10^{-11}	~10^3	1×10^{-5}
WO_3	50	50	4.2×10^{-7}		2.6×10^{-1}
WO_3 + Pd	50 + 18	50	7.4×10^{-7}	1.8	4.6×10^{-1}
CuPc + Pd	100 + 18	30	8.8×10^{-8}		5.5×10^{-2}
CuPc + Pd	200 + 18	30	2.4×10^{-10}		1.5×10^{-4}
H_2Pc	160	25	1.6×10^{-14}		1×10^{-8}
H_2Pc + Pd	160 + 27	25	5.5×10^{-9}	~10^6	3.4×10^{-3}
NiO_x + Pd	60 + 18	35	3×10^{-11}		1.9×10^{-5}

[a] Adopted volume electrical conductivity $\sigma_{Pd} = 9.26 \times 10^6$ 1/Ωm at 25 °C own results [16]

Another issue is the increase in the acoustoelectric attenuation, which is resonant. In addition, in the areas where the wave velocity receives the greatest changes, the attenuation values are also the greatest. This causes disturbances in the maintenance of stable oscillations of the SAW sensor systems. In the case of bilayer structures, there are areas of variation where the attenuating is almost constant, e.g. for $x = 10$ (see Fig. 3.18). In such configurations, where the conductivity of the second layer is about 10 times greater than the conductivity of a first layer, the phenomenon of increased attenuation in the area of large changes in the SAW velocity is much smaller, which reduces the interference of the oscillation due to the interaction of the structure with gases.

5.6.1 Summary of Experimental Research

The fifth chapter presents the results of tests of sensory properties of selected sensor structures in the SAW system, tested by the discussed measurement methods: with double-delay line, with identification of frequency modes, as well as acoustoelectric method.

The configuration and materials of the sensor structure (single, double, or multi-layer) depend on the type of gas that should be detected. Single structures in the form of thin layers of organic semiconductors (e.g. some phthalocyanins) are very well suited for detecting trace amounts of nitrogen dioxide in the air. The NO_2 detection limit in air estimated on the author's research in the paper [15] for a PbPc layer thickness of ~83 nm at ~70 °C is about 20 ppb. The use of single sensor structures in the form of thin metal layers (e.g. palladium), phthalocyanines, or metal oxides does not give satisfactory results in the detection of hydrogen, even at concentrations close to the limit of explosive in air, i.e. 4%. The frequency changes (SAW system for the dual-delay method) of such single structures do not exceed 200 Hz, and in the case of CuPc (720 nm) 350 Hz at a temperature of about 70 °C, for concentrations of 0.5–4% contained in nitrogen or air.

The use of bilayer sensor structures such as CuPc-Pd, H_2Pc-Pd, WO3-Pd, or NiO_x-Pd significantly improves hydrogen detection capabilities, even at relatively low exposure temperatures (30–40 °C) compared to traditional resistance sensors. The frequency shifts obtained are at the level of ~5 kHz for the bilayer structure with the highest thickness of hydrogen phthalocyanine (200 nm) at a concentration of 4% hydrogen in the air and an exposure temperature of ~33 °C (see Fig. 5.15). By contrast, for bilayer structures with tungsten oxide (150 nm) and palladium (10 nm) up to ~20 kHz shifts were obtained at 2% hydrogen concentration in the air, but for the temperature raised to ~100 °C (see Fig. 4.16).

For most of the studied sensor structures, a planar electrical module (see Fig. 4.11) was made in the same technological processes, enabling simultaneous monitoring of changes in electrical conductivity of the structure. The acoustoelectrical method is a source of additional information concerning the correlation of frequency shifts and electrical conductivity of sensor structures during their interaction with gases.

References

1. Anisimkin VI (1996) Russia-Italy researches on surface acoustic wave (SAW) gas sensors. In: IEEE ultrasonic symposium, pp 671–673
2. D'Amico A, Palma A, Verona E (1982) Hydrogen sensor using a palladium coated surface acoustic wale delay-line. In: IEEE ultrasonic symposium, vol 1. San Diego, pp 308–311
3. Favier F, Walter E, Zach MP, Benter T, Penner R (2001) Hydrogen sensors and switches from electrodeposited palladium mesowire arrays. Science 293:2227–2231
4. Fisher BH, Malocha DC (2010) Study of the acoustoelectric effect for SAW sensors. IEEE UFFC 57(3):698–706
5. Hechner J (1995) Formowanie warstwy ftalocyjaninu miedzi do sensorów czułych na NO_2 (Forming of copper phthalocyanine layer for NO_2 sensitive sensors). Materiały Elektroniczne ITME 23(3):24–41
6. Ho CK, Lindgren ER, Rawlinson KS, McGrath LK, Wright JL (2003) Development of a surface acoustic wave sensor for in-situ monitoring of volatile organic compounds. Sensors 3(7):236–247
7. Jakubik WP (2006) Investigations of bilayer sensor structure with copper phthalocyanine and palladium for hydrogen detection in SAW system. J Phys IV 137:95–98
8. Jakubik W (2008) Interactions of the Polyaniline and Nafion bilayer sensor structure with ammonia in a dry and wet air atmosphere. Eur Phys J Spec Top 154(1):93–96
9. Jakubik W, Krzywiecki M, Maciak E, Urbańczyk M (2012) Bi-layer nanostructures of CuPc and Pd for resistance-type and SAW-type hydrogen gas sensors. Sens Actuators B. https://doi.org/10.1016/j.snb.2012.10.056
10. Jakubik WP, Urbańczyk MW (2005) SAW hydrogen sensor with a bilayer structure based on interaction speed. Sens Actuators B 106(2):602–608

11. Jakubik WP, Urbańczyk MW, Kochowski S, Bodzenta J (2002) Bilayer structure for hydrogen detection in a surface acoustic wave sensor system. Sens Actuators B 82(2–3):265–271
12. Jakubik WP, Urbańczyk MW, Maciak E (2006) Palladium and metal-free phthalocyanine bilayer structures for hydrogen detection in the SAW sensor system based on interaction speed. IEEE Sens J 6(5):1178–1185
13. Jakubik W, Urbańczyk M, Maciak E (2007) Metal-free phthalocyanine and palladium sensor structure with a polyethylene membrane for hydrogen detection in SAW systems. Sens Actuators B 127(1):295–303
14. Jakubik WP, Urbańczyk M, Maciak E, Pustelny T (2009) Bilayer structures of NiO_x and Pd in surface acoustic wave and electrical gas sensor systems. Acta Phys Pol A 116(3):315–320
15. Jakubik W, Urbańczyk M, Opilski A (2001) Sensor properties of lead phthalocyanine in a surface acoustic wave system. Ultrasonics 39(3):227–232
16. Jakubik W, Wrotniak J, Powroźnik P (2018) Theoretical analysis of a surface acoustic wave gas sensor mechanism using electrical conductive bi-layer nanostructures. Sens Actuators B 262:947–952
17. Jakubik W, Urbańczyk M, Maciak E (2008) Detekcja wodoru z wykorzystaniem palladu oraz tlenku niklu w układzie sensorowym z akustyczną falą powierzchniową (Hydrogen detection with the use of palladium and Nikel oxide in sensor system with surface acoustic wave). Problemy Eksploatacji 4
18. Jakubik WP (2005) Hydrogen detection by single and bilayer sensor structures in Surface Acoustic Wave system. J Phys IV 129:117–120
19. Jakubik WP (2007b) Investigations of thin film structures of WO3 and WO3 with Pd for hydrogen detection in a surface acoustic wave sensor system. Thin Solid Films 515(23):8345–8350
20. Owen EA, Jones JI (1937) The palladium-hydrogen systems. Phys Soc 49:603–610
21. Penza M, Milella E, Anisimkin VI (1998) Monitoring of NH_3 gas by LB polypyrrole-based SAW sensor. Sens Actuators B 47:218–224
22. Pietrzyk M (2004) Elektryczne i akustyczne badania dwuwarstwowych struktur sensorowych (Electrical and acoustical studies of bilayer sensor structures). Master thesis, Silesian University of Technology, Gliwice
23. Tabib-Azar M, Sutapun B, Petrick R, Kazemi A (1999) Highly sensitive hydrogen sensors using palladium coated fiber optics with exposed cores and evanescent field interactions. Sens Actuators B 56:158–163
24. Urbańczyk M (2011) Sensory gazu z akustyczną falą powierzchniową (Gas sensors with surface acoustic wave). Silesian University of Technology, Gliwice
25. Urbańczyk M, Jakubik W (2010) Sposób wyznaczania koncentracji wodoru w hybrydowym czujniku wodoru z akustyczną falą powierzchniową (Method of determining hydrogen concentration in hybrid hydrogen sensor with surface acoustic wave). Patent UPRP 204036, 1 March 2010
26. Vaz AR, Salvadori MC, Cattani M (2003) Young Modulus measurement of nanostructured palladium thin films. Nanotechnology 3:177–180
27. Walter EC, Favier F, Penner R (2002) Palladium mesowire arrays for fast hydrogen sensors and hydrogen-actuated switches. Anal Chem 74:1546–1553
28. Wright JD (1989) Gas adsorption on phthalocyanines and its effects on electrical properties. Prog Surf Sci 31:1–60

Chapter 6
SAW Gas Sensors with Light-Activated Sensor Structures

Abstract This chapter presents the latest possibilities of SAW applications in gas sensors through prototype light-activated sensor structures. The nature and genesis of such structures are included, along with a qualitative model of acoustoelectric interactions, test methodology with a novel switched-channel oscillator system, and test results for thin regioregular poly-3-hexylothiophene (rrP3HT) photoconductive polymer layers in dimethylomethylophosphor (DMMP) detection (safe simulant of warfare chemical agents such as sarin). Light activation methods such as white light, low-power LED lasers, and LEDs are presented, along with ways of modulating the activation parameters by changing the intensity and wavelength of the light.

In SAW sensors, sensitivity can be regulated not only by thermal activations (as in typical resistance sensors) and geometrically by bilayer structures (with different layers 1 and 2 of surface electrical conductivity ratios) but also by optical activations, i.e. by illuminating of photoconducting sensor structures with the appropriate light stream. In this way, it is possible to create light-activated sensor structures (LASS) with increased sensitivity in SAW-type gas sensors (Fig. 6.1).

The important parameters of optical activation are the photoconductive properties of the sensor structure used (e.g. some kinds of photoconductive material) and the parameters of the light sources, i.e. the type of light source (white light, LED, low-power laser LED), the wavelength λ, light intensity I, and the angle of illumination of the structure α.

6.1 Genesis and Essence of Light-Activated Sensor Structures (LASS)

When the surface of the solid state sensor structure is brought into contact with gas molecules, the adsorption (on the surface of a thin sensing structure) or absorption (in the whole volume of the structure) takes place. For a fast detection, the adsorption processes are desirable, what can be realized mainly in a very thin structures, not

W. Jakubik, *Application of Surface Acoustic Waves in Gas Sensors*, Springer Series in Surface Sciences 69, https://doi.org/10.1007/978-3-031-89971-3_6

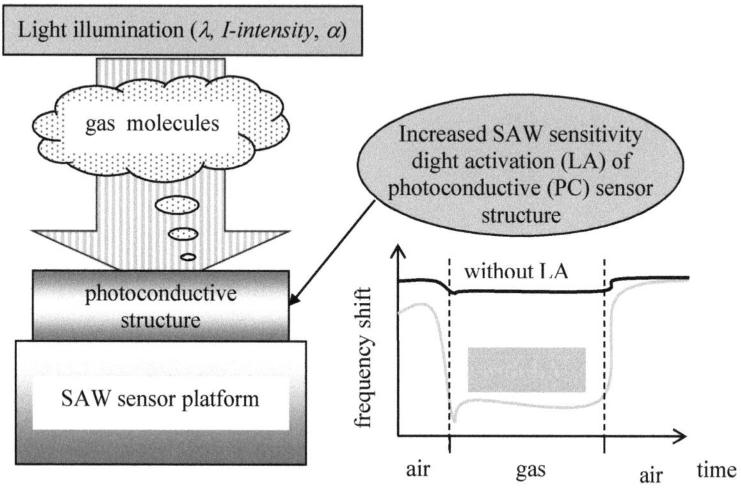

Fig. 6.1 The essence of light-activated SAW gas sensor: during the interaction of gas molecules, the photoconductive sensor structure is illuminated by a constant stream of light, which leads to a significant increase in observed frequency shifts (yellow curve with light activation) as a result of much increased sensitivity

good from the point of view of applying in a contemporary most known resistance gas detectors. The electrical resistance of a very thin layer is very often too high for a precise measuring of the changes due to interactions with gas molecules. However, light illumination often influences the adsorption/absorption rate, sometimes with a positive effect (increasing it) or even with a negative effect (decreasing it). Moreover, the time dependence of adsorption/absorption rate may also be changed by light illumination what is crucial for achieving a fast sensors. In the past, studies on the effects of light irradiation on the physical properties of the semiconducting materials (like carrier density, carrier mobility, and gas adsorption/absorption) were made mainly for common toxic gases like NO_2 in a resistance configuration [1, 2]. In SAW gas sensor technology, the additional light activation of the photoconductive sensing structures is a very promising detection method with an increasing sensitivity, initially proved with an excellent results for dimethylomethylphosphate (DMMP) detection—safe stimulant of chemical warfare agents [27, 33, 34].

In such sensor structures due to the proper light illumination, an additional electrical carriers will be generated, often with increased mobility, which is the main reason for the enhanced basic interaction rates between the sensor structure materials and the detected gas molecules. The basic interactions are the key coupling in solid-state gas sensors because without their existence the sensor cannot work properly (see Sect. 3.2). The light activation of the photoconductive sensor structures can increase the basic interactions, leading to the designing of much more efficient devices. As mentioned earlier in LASS the key parameters for achieving improved sensor parameters (i.e. the sensitivity, the response and recovery times) is the specific

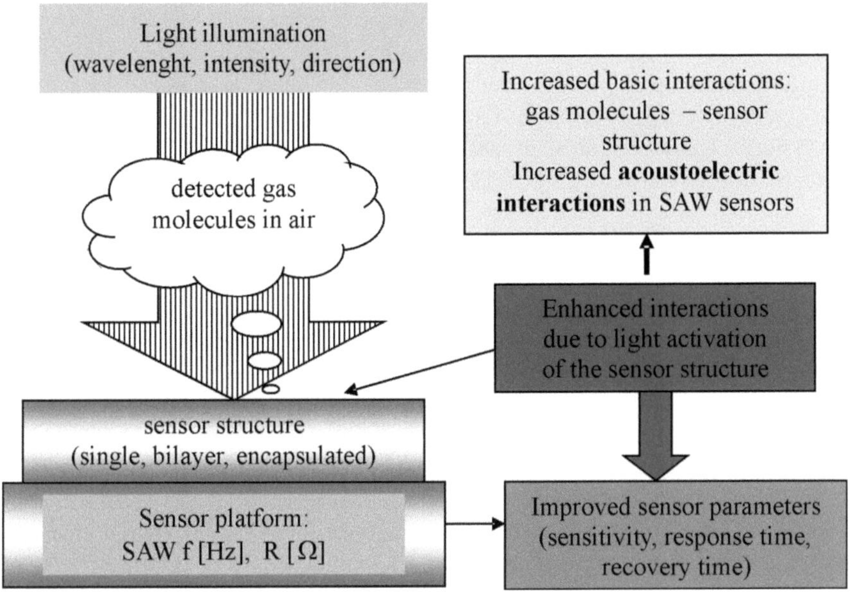

Fig. 6.2 The general idea of light-activated sensor structures (LASS) in gas sensors

matching of the light source (*i.e. the wavelength, the intensity, and the angle of illumination*) to the properly selected photoconductive sensor structure material. It can be the single structure with photoconductive polymer thin films or bilayer structures with *geometrically* increased acoustoelectric interactions, or multilayered structure with an additional encapsulated film for selectivity and stability increasing (Fig. 6.2).

6.2 Model of Acoustoelectric Interactions in LASS

In SAW gas sensors apart from the typical mass sensitivity, there is an additional possibility of an acoustoelectric (AE) interaction mechanism. As previously described (see Sect. 3.5.4), the AE effect relies on the interaction of the electric potential associated with the SAW propagated on the piezoelectric substrate with the electrical carriers in the manufactured sensor structure.

Applying an additional light activation enables the AE interactions enhancement through the moving of *structure work point* (*swp*) to the high sensitivity zone (Fig. 6.3—yellow point). The *swp* is related to the acoustoelectric parameter $\xi = \sigma_s / v_0 C_s$ (see Sect. 3.5), but now the surface conductivity of the sensor structure depends mainly on the wavelength λ of the applied light source $\sigma_s(\lambda)$ or on the light intensity $\boldsymbol{I}$ in case of a white light activation $\boldsymbol{\sigma_s(I)}$. As a result, the *swp* is shifted to the high sensitivity zone, where even very small changes in electrical surface conductivity (due to interaction with gas molecules), make great changes in SAW velocity, causing increased sensitivity of the light-activated sensor structures (Fig. 6.3). For the single metallic layers like palladium (Pd), only the mass effect can be effectively utilized because the *swp* is then placed in the non-active zone of the acoustoelectric interactions (gray point with a very high AE parameter values ~10^3–10^5 (see Fig. 6.3—the gray point). That is also the main reason for bilayer structures introduction in SAW sensor platform, such as metalophthalocyanines (MPc) with Pd, or metal oxides (like WO_3) with Pd for hydrogen detection [6–14, 16– 23].

In such bilayer configuration, the acoustoelectric fitting is much more effective and resulting sensor responses are greater and measurable (red point with AE parameter = 0.5 for WO_3 + Pd structure) [9]. However, the bilayer structures, which were fixed to the AE active zone were made only by means of the temperature increasing (planar heaters placed under the structure) or by the *technological way*, i.e. the thicknesses matching of the first (buffer layer) or the second Pd (top layer) realized only during technological processes [25]. From the other side, i.e. for the single structures of polymers, phthalocyanines, and metal oxides with a very low AE parameter ~10^{-6}–10^{-5}—yellow point, the weak AE interactions are characteristic. Consequently, the AE interactions are also not active, because, even for great changes in electrical surface conductivity (under the gas molecules influence), the significant velocity changes are not observed.

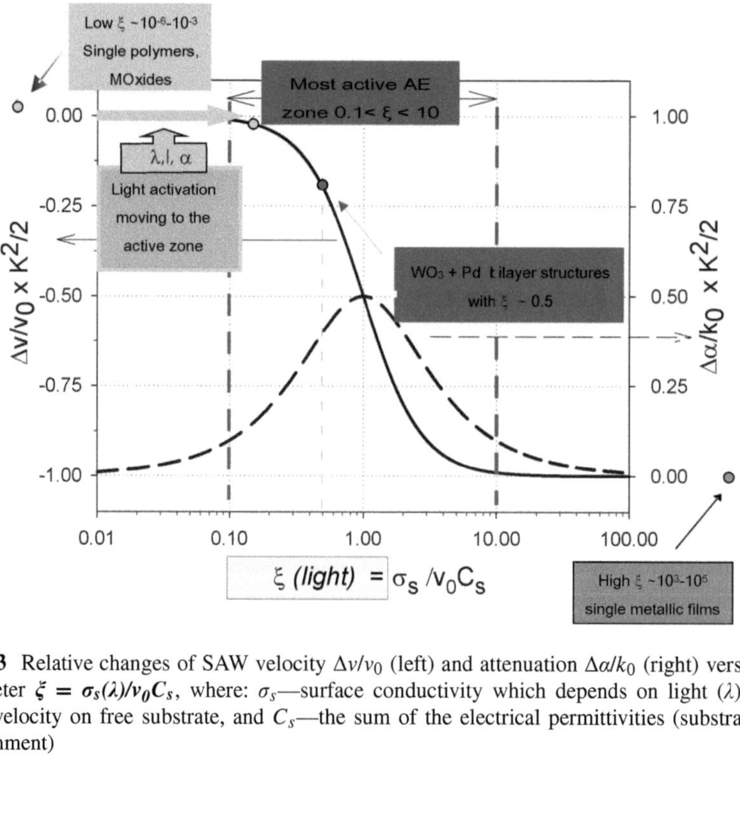

Fig. 6.3 Relative changes of SAW velocity $\Delta v/v_0$ (left) and attenuation $\Delta \alpha/k_0$ (right) versus AE parameter $\xi = \sigma_s(\lambda)/v_0C_s$, where: σ_s—surface conductivity which depends on light (λ), v_0—SAW velocity on free substrate, and C_s—the sum of the electrical permittivities (substrate and environment)

In SAW gas sensors, the optical activation relies on active shifting (under light illumination) of the *structure work point* to the active zone of the AE interactions and is the *low-temperature solution* for the initially low sensitivity of various photoconducting sensing materials, which can be effectively light-activated.

6.3 Methodology of LASS Testing

The research was carried out using the modified acoustoelectric measurement method with the possibility of light activation of the acoustic and electrical sensor modules. This method enables correlations of the obtained responses, i.e. the light-activated acoustoelectric parameter of the sensor structures can be estimated on the base of the direct electrical module resistance measurements. The method with the use of additional optical activation has been applied to a single sensor structures in a dual-delay line configuration but with the new oscillation system based on a switchable channels [27, 33].

6.3.1 Switchable Oscillator System

The known measurement systems are built in a way that they activate (by two separate generators) two SAW channels—the first is the reference element, and the second with the layer exposed to the tested gas, chemical compound, or even liquid [3, 12]. However, the use of two separate generators made of electronic components with a certain dispersion of parameters, as well as the temperature difference between two separate electronic circuits of the channels means that in many cases the system's response to the factor also depends on temperature variations and fluctuations in the supply voltage. To eliminate these inconveniences, the new proprietary *single* excitation path with a programmable phase shifter matching any sensor line from SAW to the oscillation conditions of the generator loop was developed. Due to the adopted solution of single excitation of each line separately, the generator circuit was switched between two lines with SAW during measurements. The circuit is switched by a microcontroller whose task is also to set the correct phase shift for the excited channel. In LASS investigations, a single specialized ~205 MHz generator with switchable channels of the SAW module with polymer films is developed and realized in digital technology instead of a two generator system utilized earlier (Fig. 6.4) [33].

6.3.2 Experimental and Gas Dosing System

The tested SAW module with a polymer layer in the L1 line and the free line L2 was placed in the test chamber containing the vapor inlet and outlet with the

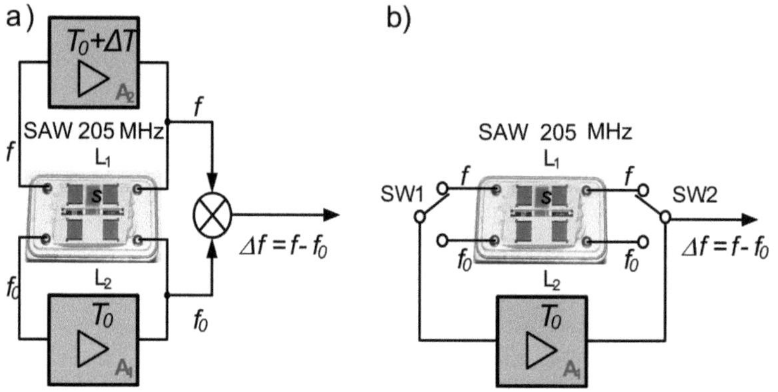

Fig. 6.4 **a** Frequently used method with a two generators, **b** single SAW generator with switchable acoustic lines; SW1 and SW2; S—sensor film; SAW 205 MHz SAW Components, Dresden, Germany

DMMP mixture, then the chamber and the module were connected to a generator with switched SW1 and SW2 paths (see Fig. 6.4b).The measuring chamber core was made of chromium-nickel steel from sheets of various thicknesses using laser cutting technology at the Institute of Medical Technology and Equipment (ITAM Zabrze, Poland). The frequency of both channels was measured with a delay between the switching of these channels at intervals of 200 ms and the time interval between consecutive readings of two samples was 5 s (Fig. 6.5).

An OWLSTONE (OVG-4) certified vapor generator was used to generate a gas mixture in dry synthetic air with a relative humidity of ~5%, which was used as a carrier gas for DMMP (Fig. 6.6a). The humidity of the flowing air can be increased and measured using the special functions of OVG-4. The concentration of DMMP vapors was determined according to the following formula [30]:

$$c = \frac{22.4 q_D}{MQ} \tag{6.1}$$

Fig. 6.5 Block diagram of the SAW experimental system with optical activation and possible four main parameters (type of light source, wavelength λ, intensity I, and the illumination angle α) for photoconducting sensor structures

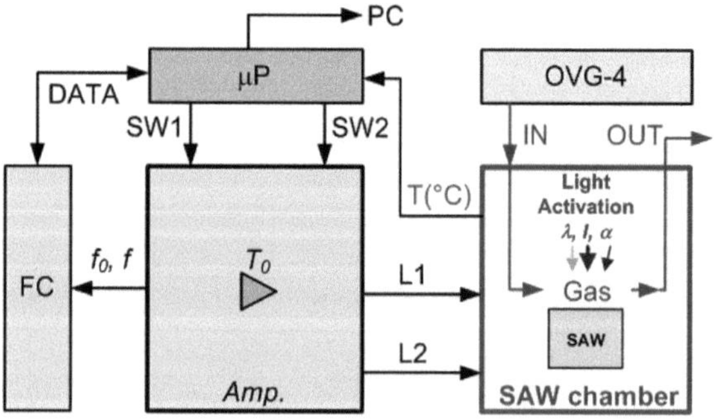

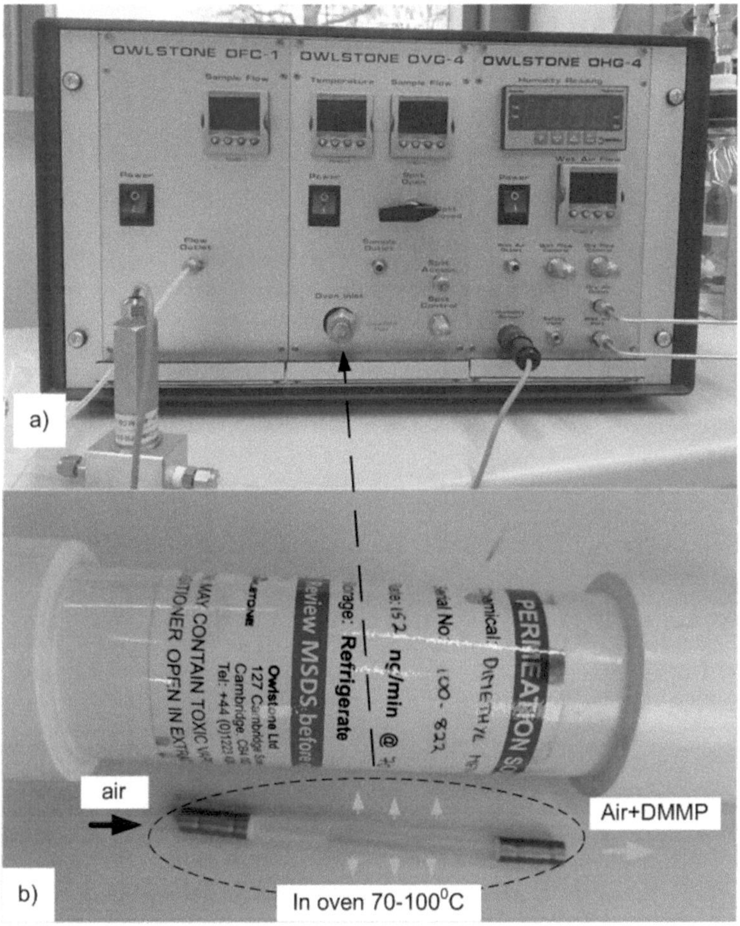

Fig. 6.6 **a** OVG-4 OWLSTONE vapor generator system, **b** certified permeation tube for DMMP vapor in ~ppm and ppb concentrations range (at various air flows and oven temperatures) in a dry synthetic air (without HC)

where c is the DMMP concentration (ppm), q_D is the certified permeation rate of the DMMP ampule (747 ng/min at 70 °C), $M = 124.08$ g/mol is the DMMP molar mass, and Q is the air flow rate (25–100 ml/min) (Fig. 6.6b). However, the permeation rate q_D can be also recalculated to the other applied temperatures (i.e. 100 °C), enabling obtaining higher concentrations.

6.3.3 Light Activation Details

Three types of light activation of the polymer films created on SAW modules were applied during measurements: a simple white light source, low-power laser LEDs, and pure LEDs. All the light sources were located directly above the SAW module with polymer film, and were mounted in the window of the chamber housing in a sealed design.

In the experiments, several light sources were used, such as white light bulbs type GY6.35 with a rated voltage of 6 V and a power of 12.5 W, common available laser LEDs (blue, green, and red with a wavelengths 405 nm, 532 nm, and 650 nm and power ~5 mW), and seven LEDs with different light wavelengths and a power of 1 W from ProLightOpto. All light source characteristics (the maximum of the wavelengths) were measured using Ocean Optics USB4000 spectrometer.

The white light source bulb circuit was powered by a voltage-regulated DC power supply in a range from 2 to 5 V, giving a light fluxes on the level from ~9 to ~137 lm (see Sect. 6.4.1) [34].

The laser LED and LED sources where current controlled (usually 100, 200, and 300 mA), giving various light fluxes in the plane of the illuminated polymer film, and the light intensity in this plane was estimated by recalculation of the measurable luxmeter value to a corresponding distance between the SAW sensor structure surface and the LED sources ~5–8 mm, respectively. The illumination area was ~2–3 mm^2.

6.4 Studies of LASS for DMMP Detection

In order to present the significant benefits of using light-activated sensor structures in gas sensors with SAW, the results for photoconductive regioregular poly-3-hexylthiophene (rrP3HT) polymer thin films with the additional light activation during the interaction with dimethyl methylphosphonate (DMMP) molecules will be presented.

6.4.1 SAW Module with Photoconductive rrP3HT Films

In studies, a commercially available quartz SAW modules with dual-delay lines operating at excitation frequency of about 205 MHz (140612-S240, SAW Components, Dresden, Germany) were used. The investigated polymer sensor layers (rrP3HT) were created in the void space of one of the delay lines. The polymer layer was deposited from a previously prepared solution, by spraying with a pistol of a nozzle thickness of 0.4 mm. Compressed synthetic air at a pressure of about 1 atm was applied as a carrier gas. The solution was prepared by dissolving ~1 mg of the polymer in 1 ml of chloroform. The distance of the nozzle from the quartz substrate

Fig. 6.7 SAW delay lines with air-sprayed rrP3HT polymer film

during the spray process was ~40 mm, whereas the deposition time was ~3 s, causing the formation of the polymer film (Fig. 6.7).

In the case of polymeric photoconductive layers (like rrP3HT), it is important to adjust the intensity and wavelength of the applied optical activation in order to obtain a compromise between increasing sensitivity and obtaining satisfactory temporal stability of such structures. In the case of white light sources, on the other hand, it is common to increase the structure temperature for higher light streams—we are dealing with optical-thermal activation (Fig. 6.8 and Table 6.1—column with resulting temperature of the structure).

6.4.2 Activation by Laser LEDs and LEDs

The experimental results for SAW sensor platform with photoconductive polymer thin film ~50 nm of rrP3HT and DMMP detection showed that the relative sensitivity increase (RSI) parameter (i.e. the comparison between the frequency shifts with optical activation of blue laser LED 405 nm and without such activation) is more than 300 times higher (Fig. 6.9) [27].

The interactions without activation are hardly visible (on the level ~4 Hz of frequency shifts), whereas the blue line obtained for activation (with blue laser LED 405 nm) presents very high frequency shift, between 690, 820, and 920 Hz for 1.5, 2, and 3 ppm of DMMP in air at low interaction temperature ~22 °C. The parameter

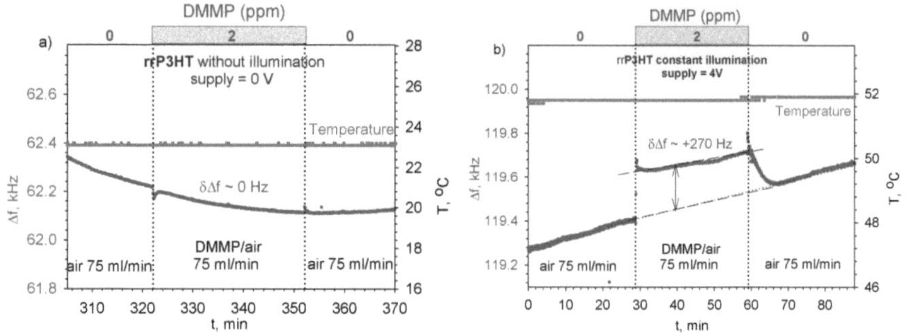

Fig. 6.8 Comparison for the interaction results **a** without activation **b** opto-thermal activation (increase of structure temperature from ~23 to ~52 °C) for rrP3HT polymer film by means of white light source

Table 6.1 White light activation results of rrP3HT film and 2 ppm of DMMP in air

Supply (V)	Estimated white light flux (lm)	Resulting temperature (°C)	Frequency shift due DMMP (Hz)
0	0	~23.2	~ 0
2	9.1	~35.5	~+130
3	24.7	~43.6	~+200
4	66.1	~52.0	~+270
5	137.3	~62.4	~+300

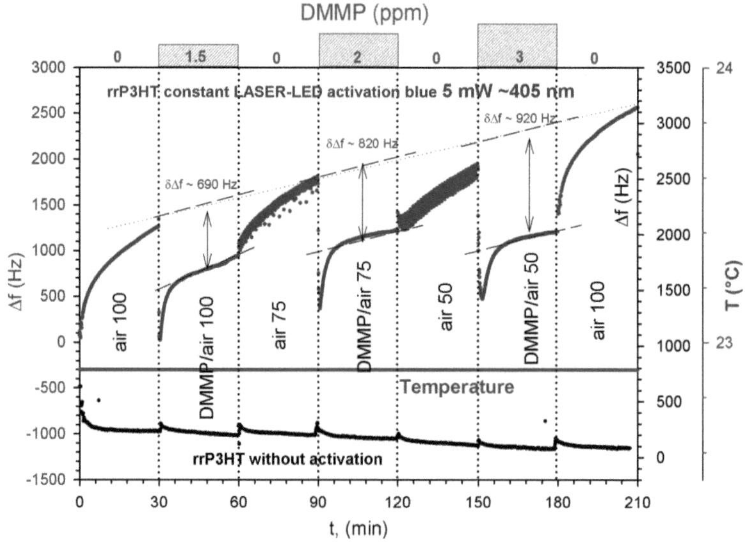

Fig. 6.9 Laser LED activation in blue range (~405 nm) of rrP3HT polymer film for 1.5, 2, and 3 ppm of DMMP concentrations in air (blue curve and left axis), and without activation (black curve and right Δf axis)

of relative sensitivity increase (RSI) can be defined for the relation of the frequency shifts as follows:

$$(\mathrm{RSI}) = \delta\Delta f\ (\text{with } LA) \big/ \delta\Delta f\ (\text{without } LA) \tag{6.2}$$

where *LA* means light activation.

The calculated values of RSI parameter are ~140, ~270, and ~310, for 1.5, 2, and 3 ppm of DMMP concentrations, respectively (see Fig. 6.9).

The given DMMP concentrations were obtained using OVG dosing system with three air flows values: 100, 75, and 50 ml/min (Eq. 6.1), what ensures very precise concentrations in measuring chamber without disturbing the sensor structure temperature. The frequency shifts $\delta\Delta f$ for a given DMMP concentration were always measured for the same flow cycle, i.e. for 100 ml/min air, then 100 ml/min air with DMMP, and similar for 75 and 50 ml/min, and the last dosing was again 100 ml/min for a faster regeneration of the structure. The above dosing sequences were utilized for all the measurements performed with the LED light-activated SAW sensor structures.

The SAW module temperature (right gray axis) is very well stabilized at the level of 22.9 °C (see Fig. 6.9). It means a considerably lower influence of the blue laser LED source on the SAW module temperature in comparison to the white light activation (see Table 6.1). However, the slow increase of the Δf signal is observed ~1500 Hz during the experiment time interval ~210 min (~7.1 Hz/min). This is mainly caused by a slow degradation of the polymer film, especially at a blue light illumination with a lower wavelength value. Such low polymer film degradation is mainly caused by photo-oxidation processes which depend on the illumination wavelength and is lower for the higher light wavelengths, i.e. for a yellow, green, or red light activation. The photo-oxidation of the rrP3HT thin films and dependence on the illumination wavelength is literature confirmed [4].

The comparison of the results in DMMP detection obtained for three lasers LED sources and the structures with rrP3HT polymer films is shown in Fig. 6.10 and the highest frequency shifts are estimated for activation with a blue light. However, the better stability is reached for the activation at green and red wavelengths.

The exemplary response of the SAW sensor structure with rrP3HT polymer film under activation with amber LED light ~590 nm (yellow curve left axis), powered by a current of 300 mA, which corresponded to the value of illumination ~57.1 × 10^3 lx in the plane of sensor structure is presented in Fig. 6.11. The estimated frequency shifts are ~220, ~260, and ~370 Hz at three DMMP concentrations in air.

The collected results for various LEDs' activation and 2 ppm of DMMP concentration in the air are presented in Fig. 6.12 for the rrP3HT sensor structures. Interesting feature is the fact that the biggest frequency shift was obtained for the yellow LED light (~585 nm) illumination with the lowest intensity ~36.2 × 10^3 lx, and the maximum parameter of RSI (relative sensitivity increase) is at the level of ~340 at the average test temperature only ~27.7 °C. A high frequency shift was observed also for a blue LED activation (~460 nm) at the lowest intensity of illumination (~30 × 10^3 lx), and here the RSI parameter was even ~420, what can be related to the intrinsic absorption of a blue light wavelengths for the rrP3HT polymer [29].

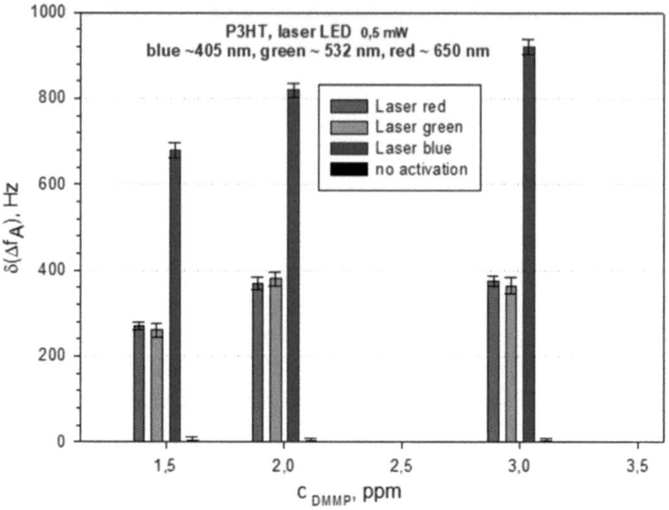

Fig. 6.10 Comparison of laser LED sources activation of rrP3HT films for the concentrations 1.5, 2, and 3 ppm of DMMP in air (the blue activation gives the biggest frequency shifts, whereas the lack of activation (black bars) are merely visible—the uncertainties were estimated on the base of a three independent measurement cycles)

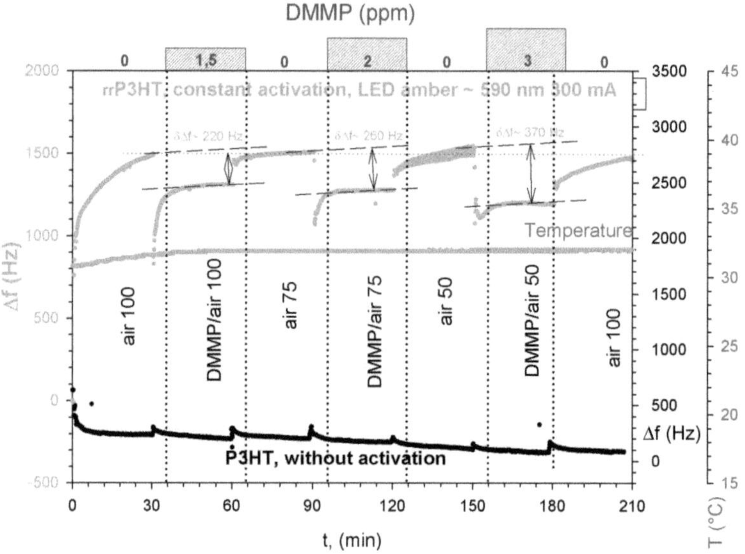

Fig. 6.11 Results for a pure LED activation (amber ~590 nm, 300 mA ~57.1 × 10^3 lx) of (rr)P3HT polymer film—the obtained relative sensitivity increase are on the level of RSI ~44 (1.5 ppm) ~87 (2 ppm) and ~123 (3 ppm)

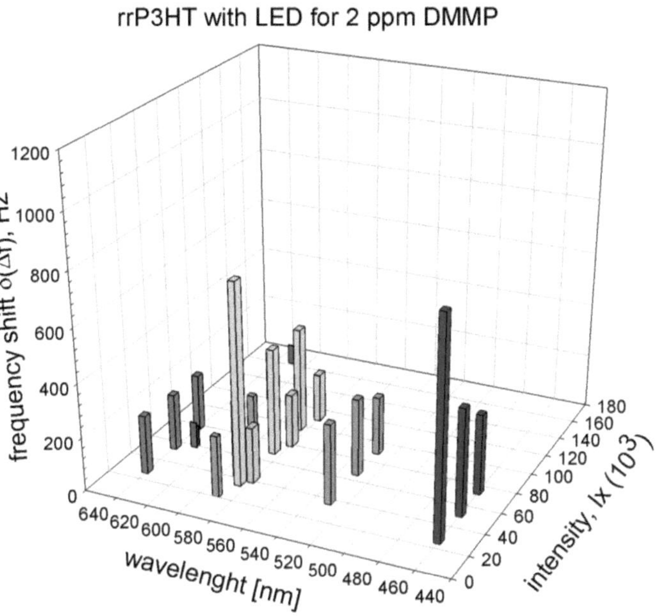

Fig. 6.12 Comparison of the results for various pure LED activation and 2 ppm of DMMP in air for rrP3HT film

6.4.3 Sensitivity on the Number of Photons

Because the frequency shifts of the tested sensor structures under the pure LED light sources were measured both, in relation to the wavelength and the intensity of the illumination, establishing the sensitivity in relation to the number of photons is interesting. Such sensitivity ($S_{\Phi N}$) can be estimated from the following formula:

$$S_{\Phi N} \propto \frac{\delta(\Delta f)}{I\lambda} \tag{6.3}$$

where, I is the intensity of the illumination and λ is the wavelength.

Figure 6.13 shows the sensitivity in relation to the number of photons for the two SAW sensor polymer structures. It is interesting to note that, at shorter wavelengths of the light activation, the sensitivity $S_{\Phi N}$ is greater for a SilPEG1.4 than for rrP3HT polymer films.

Additionally, around the yellow light wavelength, the calculated $S_{\Phi N}$ sensitivity in relation to the number of photons is almost the same for the both sensor structures. For these reasons, the yellow LED light activation with rrP3HT film at the medium light intensity has been selected to the more detailed investigations.

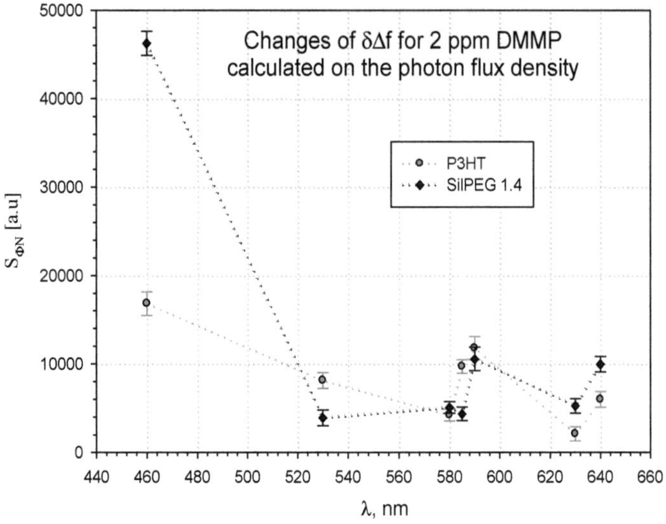

Fig. 6.13 Sensitivity calculated on the photon numbers for LED activation versus the wavelength of the light

6.4.4 Selectivity, LOD, and Response Time

The selectivity comparison for acetone, methanol, ethanol, air RH change, formaldehyde, and DMMP (recalculated for 1 ppm or 1%) is presented (Fig. 6.14a). In Fig. 6.14b, the limit of detection (LOD) (defined as a non-measurable frequency shift) is estimated on the level of ~133 ppb, and in Fig. 6.14c, the sensitivity characteristic is showed (as a minus $\delta\Delta f$) with the coefficient k_1 ~ 1013 Hz/ppm for a ppb DMMP concentrations and k_2 ~ 83 Hz/ppm at the ppm DMMP ranges. In Fig. 6.14d, the evaluation of the response times T_R is presented for 2 ppm of DMMP and for the 90% of the signal steady state, T_R is even below 5 s. However, obtaining the full steady state level requires longer time T_{Rss} ~100 s, at the low interaction temperature ~22.3 °C. Similarly, the estimated recovery time for this structure is ~240 s.

6.4.5 Repeatability and Stability of rrP3HT Films

Generally, in comparison to the white light and laser LED activation methods, the pure LED sources present a lower temperature drift of the frequency signals, lower film degradation at a longer light wavelengths, and greater frequency shifts for the same DMMP concentrations. The lower degradation (in comparison to the blue laser LED ~405 nm—see Fig. 6.9) is present for yellow LED activation (~585 nm) (Fig. 6.10). The stability test for three same DMMP dosing cycles (with the estimated frequency coefficient increase ~0.92 Hz/min—much lower than for the blue laser activation

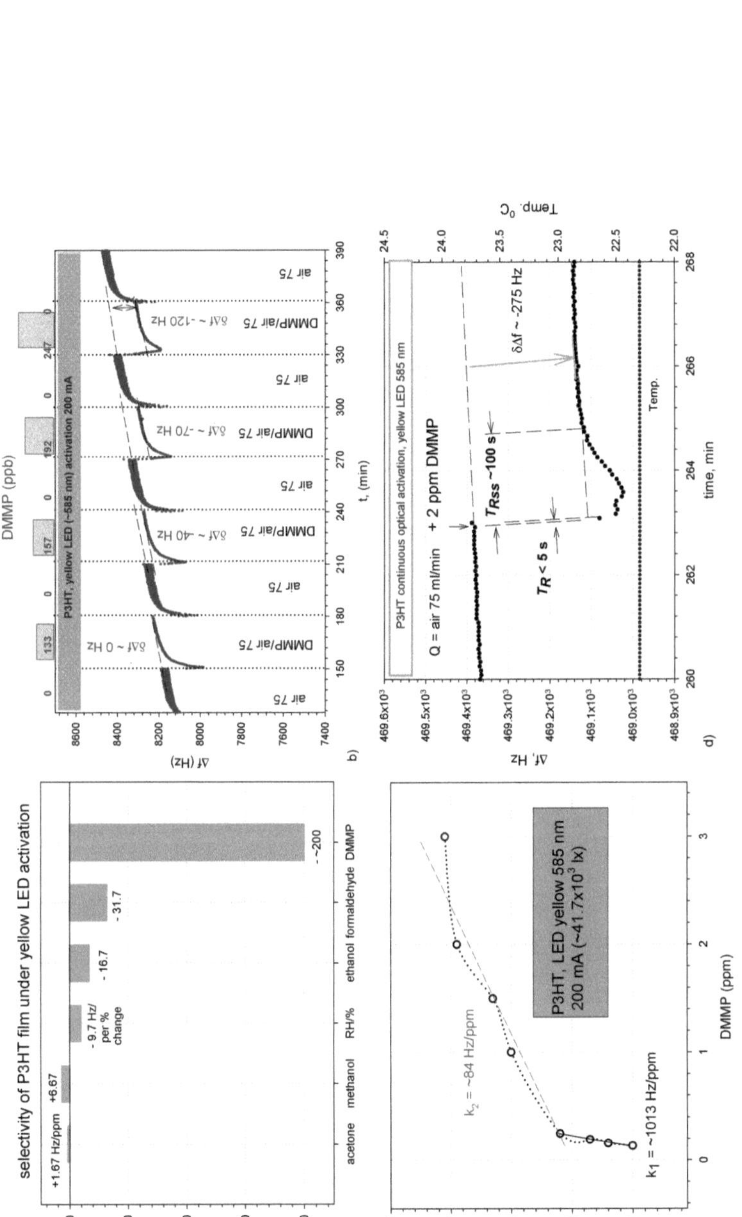

Fig. 6.14 **a** Selectivity comparison for acetone, methanol, ethanol, air RH change, formaldehyde, and DMMP (for 1 ppm or 1%), **b** LOD test ~133 ppb, **c** the sensitivity characteristic, **d** response time evaluation for the SAW sensor structure with P3HT film with yellow LED activation (585 nm) and average utilized intensity ~42 × 10^3 lx (200 mA) [27]

~7.1 Hz/min) is shown in Fig. 6.15a, whereas the repeatability (on the base of (a) results which are collected in Table 6.2)—in Fig. 6.15b.

6.5 Acoustoelectrical Method in LASS

6.5.1 *Electrical Planar Transducer*

In order to confirm the existence of the acoustoelectric interactions, a planar electrical transducer coated with a photoconductive polymer layer has been made and tested (Fig. 6.16) (see Sect. 4.6). The rrP3HT polymer film in this transducer was manufactured in the same spray-coating technology processes as the SAW module. The important parameters of the transducer are N is the number of the transducer electrodes pairs (=12), p is the electrode width and distance between electrodes (~0.5 mm), l is the width of the electrode system cover (~12.4 mm).

Basing on these transducer parameters and the results of directly measured electrical resistance $R_s(light)$ of the sample after light activation, it is possible to determine the surface electric conductivity in function of the utilized light sources [32]:

$$\sigma_{s(light)} = \frac{1}{R_{s(light)}} \frac{p}{(2N - 1)l} \tag{6.4}$$

and then estimate the acoustoelectric parameter ξ *(light)* of the SAW sensor structure with rrP3HT polymer layer after the light activation [15]:

$$\xi_{light} = \frac{\sigma_{s(light)}}{v_0 \cdot C_s} \tag{6.5}$$

where $v_0 = 3158$ m/s is the SAW velocity in ST quartz substrate, $C_S = 0.5$ pF/cm is the capacity on the length unit of the SAW module, and $\sigma_{s(light)}$ is the electrical surface conductivity of the polymer film calculated from the sample resistance $R_{s(light)}$ after the light activation (Eq. 6.4).

The estimated value of the AE parameter ξ_{light} shows how effective are the acoustoelectric interactions in case of the light activation process, because the ξ values can be determined directly after film deposition (on the base of R_{s0} value) and then after various light activation processes (on the base of $R_{s(light)}$ values) (Fig. 6.17). The great SAW sensor structures sensitivity increase after the light activation can be explained in the category of AE parameter shifting to the high sensitivity region (see Fig. 6.17—activation in a yellow and blue LED).

The shifting of the *structure work point* for AE interactions has been confirmed by measurements in a planar resistance sensor platform, which enable independently determining of the AE parameter for SAW platform due to light activation. The resistance sensor platform allows comparison of the obtained AE sensitivity under the light illumination and without such activation. The points on the AE sensitivity

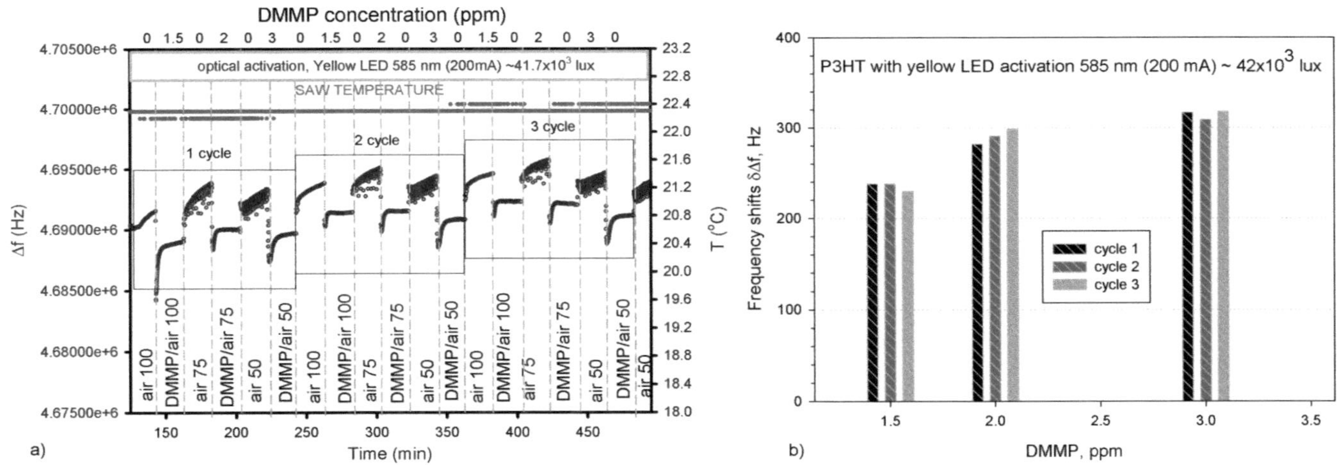

Fig. 6.15 **a** Stability test for three same DMMP dosing cycles (estimated frequency coefficient increase ~0.92 Hz/min—as measure of instability), **b** repeatability on the base of **a** results collected in Table 6.2, for rrP3HT polymer structure with yellow LED activation [27]

Table 6.2 Estimated frequency shifts for rrP3HT film with yellow LED activation and three measuring cycles (on the base of Fig. 6.15a)

DMMP concentration (ppm)	Frequency shifts (Hz) Cycle no		
	1	2	3
1.5	238	238	230
2	282	291	299
3	317	309	318

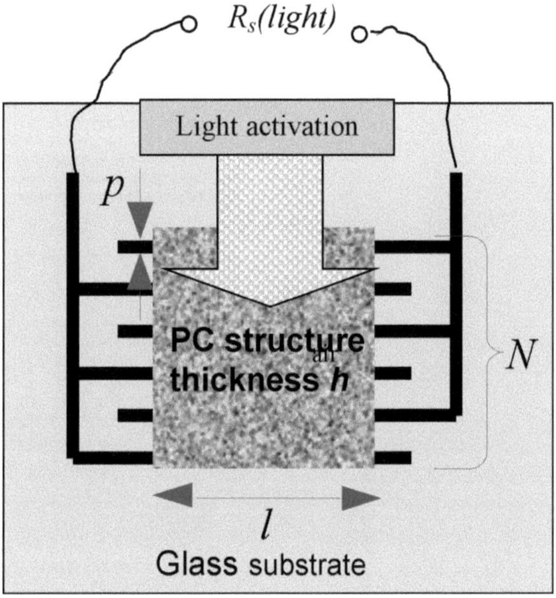

Fig. 6.16 The scheme of electrical resistive transducer for electrical resistance $R_s(light)$ measuring of photoconductive (PC) structure after light activation

curve (see Fig. 6.17), observed for the investigated ~50 nm rrP3HT polymer sensor structure, present the increasing sensitivity due to application of various optical illumination. The highest AE sensitivity is detected for the blue and yellow LED light activations (Fig. 6.17—the yellow and blue points) with perpendicular illuminations from the top at average light intensities. The higher intensities occurs slow degradations of the polymer film in ~dozen hours [27].

6.5.2 Explanation of Sensitivity Increase

The observed sensitivity increase of the light activated SAW polymer structures for DMMP detection can be interpreted in two categories. Firstly, the base molecular adsorption rate of the DMMP molecules is enhanced, what is especially probable

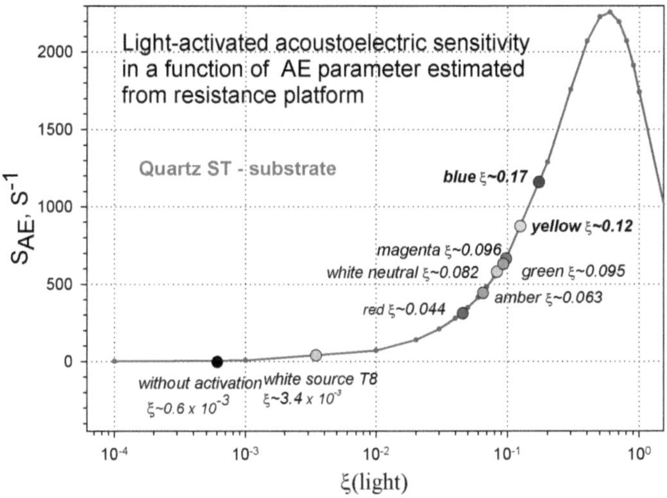

Fig. 6.17 Light-activated AE sensitivity S_{AE} for photoconductive (rr)P3HT polymer film on quartz ST substrate as function of AE parameter (estimated on the base of resistance platform with optical activation)—the *greatest sensitivity* observed for yellow and blue LED light as in SAW investigations

for activation in a blue illumination range, where the absorption for rrP3HT polymer films is observed [29]. This photoactivation of the rrP3HT layers in air upon exposure to irradiation close to the UV is caused to the photogeneration of charge carriers on the polymer chain. This photogeneration takes place via the absorption of a quantum of energy (hv) to induce charge separation on neighboring repeat units of the polymer and is a well-known phenomenon [31].

Secondly, the increased acoustoelectric (AE) interaction between SAW and sensor polymer structure are possible, because the significant increase of electrical conductivity of the polymer structure is observed after the light illumination, resulting in a shifting of the sensor *structure work point* to the active acoustoelectric interaction zone [13, 15, 28]. This assumption has been further experimentally confirmed by an electrical planar method using resistive transducer (see Fig. 6.16). The initial (after deposition) sample resistance R_{s0} ~19.2 MΩ of rrP3HT decreased significantly after a white light illumination to the R_{slight} ~3.35 MΩ value and later only slightly increased (~+50 kΩ) due to the 2 ppm DMMP adsorption (Fig. 6.18 (not in scale)). The next utilized LED activation processes (red, yellow, and blue) gradually decrease the sample resistance with the smallest value at the blue LED illumination (~0.067 MΩ). It means that the light activation processes of the photoconductive polymer thin films significantly increase their electrical surface conductivity (see Eq. 6.4), by generating additional charge carriers. In SAW sensor configuration the acoustoelectric interactions (between the SAW electric potential and the enriched charge carriers in the polymer film) are possible, especially for a yellow and blue LED illumination.

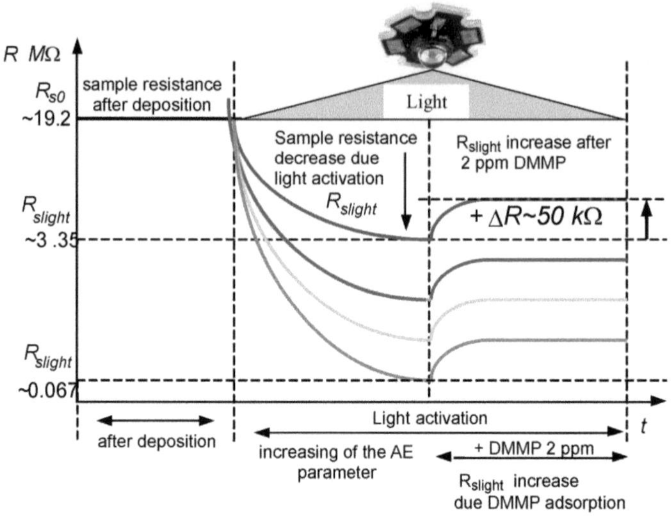

Fig. 6.18 The scheme of the electrical resistance changes (not in scale) for P3HT polymer sample (on the base of resistive transducer results—see Table 6.3) after light activations: white light (black curve) and red, yellow, and blue LEDs, where: R_{s0} is the sample resistance after deposition, R_{slight} is the sample resistance after light activation, increase in R_{slight} ~+50 kΩ after 2 ppm of DMMP

It is confirmed (see Fig. 6.17), where the increase of acoustoelectric sensitivity S_{AE} is presented after different optical activations (white light and LEDs) in function of the acoustoelectric parameter ξ_{light} (calculated from Eq. 6.5 on the base of the resistive planar transducer results and Eq. 6.4) for the rrP3HT polymer film on a quartz substrate. The greatest sensitivity shifts are obtained for a blue and yellow LED light activation, what is in accordance with the obtained SAW results. Additionally, the activation by a yellow LED wavelength should be interpreted in the second category of AE interactions, because there is no literature confirmation of the increased absorption efficiency of the rrP3HT polymer films in such wavelength range. The results of the rrP3HT polymer film parameters are collected in Table 6.3.

Table 6.3 Parameters of rrP3HT estimated from resistive transducer sample

Light activation	R_{s0}, $R_{s(light)}$, MΩ	σ_{s0}, $\sigma_{s(light)}$ $\times 10^{-9}$, Ω^{-1}	ξ_0, ξ_{light} $\times 10^{-3}$
No	19.2	0.094	0.6
White light	3.35	0.54	3.4
Red LED	0.66	2.73	17.3
Yellow LED	0.48	3.75	23.7
Blue LED	0.067	26.9	170

R_{s0}, $R_{s(light)}$—experimental results of the sample resistance, without and under the light activation, respectively, σ_{s0}, $\sigma_{s(light)}$—calculated sheet conductivity values (Eq. 6.4), ξ_0, ξ_{light}—calculated AE parameter values (Eq. 6.5) for the average light intensity and long time (~10 h) illumination [27]

6.6 Summary for LASS

In SAW gas sensor technology, the additional light activation of the polymeric sensor structure for DMMP detection has been not yet utilized and for selected photoconductive polymer films (like rrP3HT) is a novel solution. The development of light-activated polymer sensor structures in SAW configuration, is a considerable contribution to a progress in SAW gas sensing technology through the more effective use of the acoustoelectric sensitivity.

A suitable light activation of the photoconductive polymer sensor structures in SAW configurations leads to a much higher device frequency shifts, and consequently to a much higher sensitivity in gas detection. The observed much increasing of frequency shifts can be interpreted in the category of acoustoelectric interactions, i.e. the shifting of the structure work point to the high sensitivity zone due to the light activating process (Fig. 6.19). In case of a constant or slowly changing humidity levels, such structures are a simple way for developing fast and low-temperature gas sensors, which can be comparable with the current existing devices.

The additional light activation creates a new category of the *active sensor structures* in the field of SAW gas sensor platforms.

The most important parameters of the light-activated sensor structures are the photoconductive properties of the sensor material, the kind of light source (white

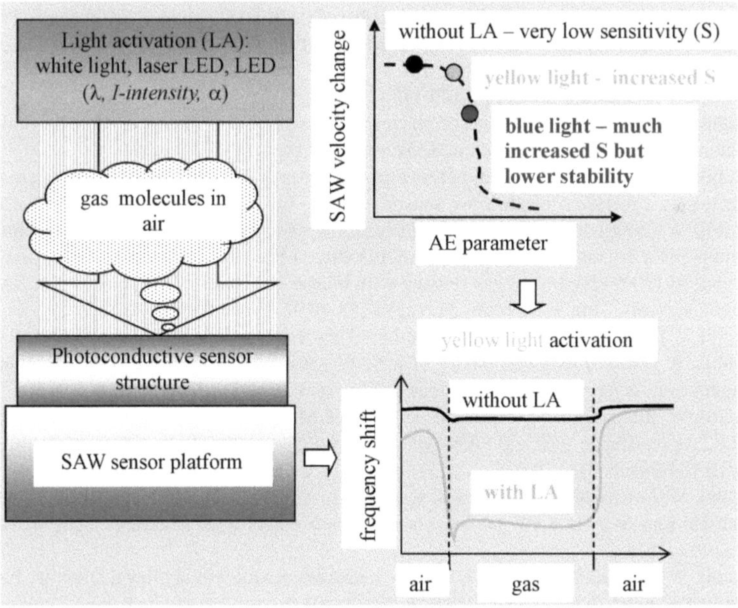

Fig. 6.19 The much increased frequency shifts in SAW gas sensors due to light activating process explained in category of acoustoelectric interactions of photoconductive sensor structure, LA—light activation

light, LEDs, low-power laser LEDs) the light wavelength, intensity, and direction of sensor structure illumination.

The last condition (i.e. the illumination angle) is important when using bilayer structures for which the second layer is opaque. Light activations can then be carried out by side illumination or from below via a properly designed fiber optic system. However, this type of activation has not yet been used in SAW-type sensors and may in the future be a new solution for designing devices with much better sensitivity parameters and shorter response and regeneration times.

References

1. Comini E, Cristalli A, Faglia G, Sberveglieri G (2000) Light enhanced gas sensing properties of indium oxide and tin dioxide sensors. Sens Actuators B 65:260–263
2. Comini E, Faglia G, Sberveglieri G (2001) UV light activation of tin oxide thin films for NO_2 sensing at low temperatures. Sens Actuators B 78:73–77
3. Du J, Harding G, Ogilvy J, Dencher P, Lake M (1996) A study of Love-wave acoustic sensors. Sens Actuators A 56:211–219
4. Hintz H, Sessler C, Peisert H, Egelhaaf HJ, Chasse T (2012) Wavelength-dependent pathways of poly-3-hexylthiophene photooxidation. Chem Mater 24:2739–2743
5. Jakubik W (2012) Zastosowanie akustycznych fal powierzchniowych w czujnikach gazowych (Application of surface acoustic waves in gas sensors). Wydawnictwo Politechniki Śląskiej, Silesian University of Technology, Monograph no. 413
6. Jakubik W (2012) Theory of SAW gas sensor based on bi-layer conductivity changes. Procedia Eng 1287–1290. Proc Eurosensors XXVI, Kraków, Poland
7. Jakubik WP (2005) Hydrogen detection by single and bilayer sensor structures in Surface Acoustic Wave system. J Phys IV 129:117–120
8. Jakubik WP (2006) Investigations of bilayer sensor structure with copper phthalocyanine and palladium for hydrogen detection in SAW system. J Phys IV 137:95–98
9. Jakubik WP (2007) Investigations of thin film structures of WO_3 and WO_3 with Pd for hydrogen detection in a surface acoustic wave sensor system. Thin Solid Films 515(23):8345–8350
10. Jakubik W (2008) Interactions of the Polyaniline and Nafion bilayer sensor structure with ammonia in a dry and wet air atmosphere. Eur Phys J Spec Top 154(1):93–96
11. Jakubik W (2009) Hydrogen gas-sensing with bilayer structures of WO_3 and Pd in SAW and electric systems. Thin Solid Films 517(22):6188–6191
12. Jakubik W (2011) Surface acoustic wave-based gas sensors. Thin Solid Films 520(3):986–993
13. Jakubik W (2014) Elemental theory of a SAW gas sensor based on electrical conductivity changes in bi-layer nanostructures. Sens Actuators B 203:511–516
14. Jakubik W, Krzywiecki M, Maciak E, Urbańczyk M (2012) Bi-layer nanostructures of CuPc and Pd for resistance-type and SAW-type hydrogen gas sensors. Sens Actuators B. https://doi.org/10.1016/j.snb.2012.10.056
15. Jakubik W, Powroźnik P, Wrotniak J, Krzywiecki M (2016) Theoretical analysis of acousto-electrical sensitivity in SAW gas sensors with single and bi-layer structures. Sens Actuators B 236:1069–1074
16. Jakubik WP, Urbańczyk MW (2005) SAW hydrogen sensor with a bilayer structure based on interaction speed. Sens Actuators B 106(2):602–608
17. Jakubik WP, Urbańczyk MW, Bethanabotla V, Cular S (2006) Acoustoelectric effect in Surface Acoustic Wave gas sensors (SAW) with metal-free phthalocyanine and palladium sensing bi-layers. ECS Trans 1(19):1–9

18. Jakubik WP, Urbańczyk MW, Bodzenta J, Pietrzyk MA (2005) Investigations on the resistance of the metal-free phthalocyanine and palladium bilayer sensor structure influenced by hydrogen. Sens Actuator B 105(2):340–345
19. Jakubik WP, Urbańczyk MW, Kochowski S, Bodzenta J (2002) Bilayer structure for hydrogen detection in a surface acoustic wave sensor system. Sens Actuators B 82(2–3):265–271
20. Jakubik WP, Urbańczyk MW, Kochowski S, Bodzenta J (2003) Palladium and phthalocyanine bilayer films for hydrogen detection in a surface acoustic wave sensor system. Sens Actuators B 96(1–2):321–328
21. Jakubik W, Urbańczyk M, Maciak E (2008) Detekcja wodoru z wykorzystaniem palladu oraz tlenku niklu w układzie sensorowym z akustyczną falą powierzchniową (Hydrogen detection with the use of palladium and nickel oxide in sensor system with surface acoustic wave). Problemy Eksploatacji 4
22. Jakubik WP, Urbańczyk MW, Maciak E (2006) Palladium and metal-free phthalocyanine bilayer structures for hydrogen detection in the SAW sensor system based on interaction speed. IEEE Sens J 6(5):1178–1185
23. Jakubik W, Urbańczyk M, Maciak E (2007) Metal-free phthalocyanine and palladium sensor structure with a polyethylene membrane for hydrogen detection in SAW systems. Sens Actuators B 127(1):295–303
24. Jakubik W, Urbańczyk M, Maciak E, Krzywiecki M (2011) Bi-layer nano-structures of CuPc and Pd in SAW and resistance hydrogen sensor. Procedia Eng 25:252–255
25. Jakubik WP, Urbańczyk M, Maciak E, Pustelny T (2008) Surface acoustic wave hydrogen gas sensor based on layered structure of palladium/metal-free phthalocyanine. Bull Pol Acad Sci Tech 56(2):133–138
26. Jakubik WP, Urbańczyk M, Maciak E, Pustelny T (2009) Bilayer structures of NiO_x and Pd in surface acoustic wave and electrical gas sensor systems. Acta Phys Pol A 116(3):315–320
27. Jakubik W, Wrotniak J, Kaźmierczak-Bałata A, Stolarczyk A, Powroźnik P (2023) Light-activated SAW sensor structures with photoconductive polymer films for DMMP detection. Sens Actuators B 384:133597. https://doi.org/10.1016/j.snb.2023.133597
28. Jakubik W, Wrotniak J, Powroźnik P (2018) Theoretical analysis of a surface acoustic wave gas sensor mechanism using electrical conductive bi-layer nanostructures. Sens Actuators B 262:947–952
29. Kępska K, Jarosz T, Januszkiewicz-Kaleniak A, Domagała W, Łapkowski M, Stolarczyk A (2018) Spectroelectrochemistry of poly(3-hexylthiophenes) in solution. Chem Zvesti 72(1):251–259
30. Powroźnik P, Jakubik W, Stolarczyk A, Kazmierczak-Balata A, Wrotniak J, Jarosz T (2020) Study of light-activated regioregular poly(3-hexyltiophene) photoconductive polymer sensing properties in nerve agent simulant (DMMP) detection. Sensors 20(2):491
31. Vardeny Z, Ehrenfreund E, Brafman O, Nowak M, Schaffer H, Heeger A, Wudl F (1986) Photogeneration of confined soliton pairs (bipolarons) in polythiophene. Phys Rev Lett 56:671
32. Wohltjen H, Barger W, Snow A, Jarvis NL (1985) A vapour-sensitive chemoresistor fabricated with planar microelectrodes and a Langmiur-Blodgett organic semiconductor film. IEEE Trans ED 32(7):1170–1174
33. Wrotniak J (2022) Akustyczne fale powierzchniowe w badaniach własności sensorowych cienkich warstw wybranych związków polimerowych z wykorzystaniem dodatkowych metod aktywacji (Surface Acoustic Waves in studies of sensor properties of selected polymer compounds thin films with the use of additional activation methods). Dissertation, Silesian University of Technology, Gliwice
34. Wrotniak J, Jakubik W, Powroźnik P, Stolarczyk A, Magnuski M (2018) Akustyczne badania polimeru typu RR-P3HT do wykrywania DMMP w powietrzu (Acoustical studies of polymer RR-P3HT type for the determination of DMMP in air). Electrotech Rev 94(6):70–73

Chapter 7
Summing-Up

Abstract The summary (this chapter) highlights the author's original contribution to the development of gas sensors with SAW. The prospects of applications and further research taking into account the latest solutions based on light-activated photoconductive polymer structures are also indicated.

Acoustic surface waves have been used in sensor technology for many years. The essence of SAW gas sensors is a sensor element, usually in the form of a thin layer, which is sensitive (changes its parameters) in the presence of even trace amounts of certain vapor or gas molecules. SAW sensor systems are sensitive not only to changes in mass but also to changes in the electrical conductivity of the used structure. The coupling between changes in electrical conductivity and SAW occurs as a result of acoustoelectric interactions. Such interactions can be enhanced for sensor elements in the form of bilayer semiconductor–metal structures. The use of single thin layers (e.g. palladium or WO_3) gives much weaker results with respect to detection, e.g. hydrogen, than the use of bilayer structures such as phthalocyanine-palladium or metal oxide-palladium. Bilayer structures of this kind open up new possibilities in gas sensors with SAW. Therefore, not only the sensor material is important but also its proper layer configuration.

SAWs make it possible to study the sensor properties of different materials in the form of thin layers or layer structures. However, their thickness must be many times smaller than the length of the propagating surface wave. For much larger thicknesses or materials with a high viscosity ratio, wave energy will radiate to the sensor structure. In these cases, there are problems with the excitation and maintenance of the oscillation, e.g. in the dual-delay line method. An undoubted advantage of using SAW in sensor technology is the ability to test the sensor properties of thin layers or entire structures with very high resistance. Electrical methods are then almost impossible to use due to high noise levels and low measurement accuracy. In SAW gas sensors, the resistance of the structure is practically irrelevant. There may be no acoustoelectric interactions, as a result of the mismatch to an active area with high sensitivity. Mass-elastic interactions, on the other hand, occur almost always, regardless of the electrical parameters of the sensor materials used.

W. Jakubik, *Application of Surface Acoustic Waves in Gas Sensors*, Springer Series in Surface Sciences 69, https://doi.org/10.1007/978-3-031-89971-3_7

The author's original achievements regarding the use of SAW in gas sensors presented in the monograph are:

1. Introduction and use of *bilayer sensor structures* in SAW gas sensors and the possibility of using acoustoelectric interactions for such systems [5, 11, 12, 13, 14, 15, 16, 21]
2. Confirmation of acoustoelectric interactions for bilayer sensor structures by simultaneous measurements of changes in electrical conductivity due to interaction with hydrogen—very good acoustoelectric correlation for structures like MPc-Pd and WO_3-Pd [2, 3, 7–9, 12, 17, 20, 22].
3. Using a new detection method for H_2Pc-Pd bilayer structures. Different rates of differential frequency change were obtained depending on the hydrogen concentration [10, 18].
4. Use of additional polyethylene (PE) layers for H_2Pc-Pd layer structures to reduce water vapor impact—single and double prototype PE layers—18-fold reduction of water vapor impact by RH ~50% [19].
5. Application of additional protective layers of polymethylopentan (TPX), for bilayer structures of type WO_3-Pd, to reduce the influence of water vapor and increase the selectivity towards ammonia.
6. Bilayer nafion-polyaniline structures, giving the possibility to detect trace amounts of ammonia (50 ppm) in humid atmosphere RH ~50% and at relatively low interaction temperatures ~30 °C, compared to traditional resistance sensors [1].
7. Analysis, using perturbation theory, of the effect of bilayer sensor structures on SAW propagation in piezoelectric substrates. Obtaining expressions describing the relative changes in attenuation and velocity in the electrical conductivity of each layer [4, 6, 23].
8. Use of active sensor structures (with appropriately selected light activation) in the detection of DMMP (simulant of warfare agents, e.g. sarin) through photoconductive polymer layers [24].

Layered sensor structures, e.g. semiconductor–metal and semiconductor–metal-selective membrane, in SAW sensors represent an important contribution to the development of gas sensing technology. Bilayer structures, for which a theoretical analysis was carried out for the first time, create new possibilities for exploiting their high acoustoelectric sensitivity. The design parameter of such structures, i.e. the ratio of the electrical surface conductivity of the thin layers of the materials used, determines the operating points for which the greatest changes in SAW velocity, and thus the greatest sensitivity, can be obtained. In small ranges, the working points of the made sensor structures can be changed by increasing their working temperature. The larger the design parameter of the structures, the more the range of the largest wave speed changes is shifted towards smaller values of the acoustoelectric parameter, which is a function of the electrical conductivity of the first layer. The maximum change in SAW speed corresponds to the inverse of the design parameter of the sensor structure.

The very high sensitivity of sensor systems with SAW is a consequence of the high density of wave energy in the surface area. This density is increased in systems with Love wave, by making additional waveguide layers. Such systems create great possibilities in the detection of trace amounts of ions or enzymes present in the liquid phase. They form the basis for the currently intensively developed group of biosensors that are used in biology and medicine. However, it should be borne in mind that even the highest sensitivity of the detection method will not result in the correct functioning of the sensor without the appropriate sensor element. Nevertheless, the very high sensitivity of the method allows detection of certain substances, even with a sensor element with poor sorption properties.

The sensitivity of SAW sensor systems based on acoustoelectric interactions can be varied by matching them to the active region of their characteristics, where the wave speed changes are particularly large. The electrical conductivity of the sensor structure must be affected. This creates additional possibilities for the design of active sensor structures whose electrical conductivity would be altered by additional optical illumination with appropriate wavelength. The sensitivity of such sensor structures could be adapted to the range of concentrations of specific vapors or gases. Active sensor structures of this type, have been used in practical implementations of gas sensors with SAW with light-activated sensor structures of pthotoconductive rrP3HT polymer films in DMMP detection.

The monograph presents the most important issues related to the application of surface acoustic waves (SAW) in gas sensors.

The work contains general characteristics of SAW sensors, taking into account the essence of the possibility of using this type of waves in gas sensors, as well as the basic method of their generation and detection in piezoelectric substrates by means of interdigital transducers.

An important part of the work are the general models of the SAW sensor operation along with their mathematical description. The general model was based on three basic couplings found in gas sensors: gas molecules-sensor structure, then sensor structure—SAW propagation and measurement of altered SAW parameters through a suitably sensitive detection system. The mathematical description takes into account the influence of the most important parameters of the layers and sensor structures on the propagation of the SAW. The description adopts a typical division into the mass effect, resulting from the mechanical parameters of the sensor structure, and the acoustoelectric effect, which is a consequence of its electrical parameters. For the first time, a theoretical description of the acoustoelectrical effect for bilayer sensor structures has been presented.

The book describes important features of the most commonly used measurement methods, including a measuring system with a double-delay line. This system compensates for the influence of some undesirable external factors on the measurement results, such as small changes in temperature or pressure.

The original results of the sensor properties studies of selected single, bilayered structures, as well as prototype multi-layer structures, in which additional layers perform limiting functions, e.g. the influence of water vapor, were presented. Most attention was paid to semiconductor–metal bilayer structures, such as phthalocyanine-palladium or tungsten oxide-palladium, as they are the most promising in terms of practical applications in SAW hydrogen sensors.

The summary highlights the author's original contribution to the development of gas sensors with SAW. The prospects of applications and further research taking into account the latest solutions based on light-activated photoconductive polymer structures are also indicated.

References

1. Jakubik W (2008) Interactions of the polyaniline and Nafion bilayer sensor structure with ammonia in a dry and wet air atmosphere. Eur Phys J Spec Top 154(1):93–96
2. Jakubik W (2009) Hydrogen gas-sensing with bilayer structures of WO_3 and Pd in SAW and electric systems. Thin Solid Films 517(22):6188–6191
3. Jakubik W (2011) Surface acoustic wave-based gas sensors. Thin Solid Films 520(3):986–993
4. Jakubik W (2012) Theory of SAW gas sensor based on bi-layer conductivity changes. Procedia Eng 1287–1290. Proc Eurosensors XXVI, Kraków, Poland
5. Jakubik W (2012) Zastosowanie akustycznych fal powierzchniowych w czujnikach gazowych (Application of surface acoustic waves in gas sensors). Wydawnictwo Politechniki Śląskiej, Silesian University of Technology Monograph no. 413
6. Jakubik W (2014) Elemental theory of a SAW gas sensor based on electrical conductivity changes in bi-layer nanostructures. Sens Actuators B 203:511–516
7. Jakubik WP (2005) Hydrogen detection by single and bilayer sensor structures in Surface Acoustic Wave system. J Phys IV 129:117–120
8. Jakubik WP (2006) Investigations of bilayer sensor structure with copper phthalocyanine and palladium for hydrogen detection in SAW system. J Phys IV 137:95–98
9. Jakubik WP (2007) Investigations of thin film structures of WO_3 and WO_3 with Pd for hydrogen detection in a surface acoustic wave sensor system. Thin Solid Films 515(23):8345–8350
10. Jakubik WP, Urbańczyk MW (2005) SAW hydrogen sensor with a bilayer structure based on interaction speed. Sens Actuators B 106(2):602–608
11. Jakubik W, Urbańczyk M, Maciak E (2008) Detekcja wodoru z wykorzystaniem palladu oraz tlenku niklu w układzie sensorowym z akustyczną falą powierzchniową (Hydrogen detection with the use of palladium and Nickel oxide in sensor system with surface acoustic wave). Problemy Eksploatacji 4
12. Jakubik W, Krzywiecki M, Maciak E, Urbańczyk M (2012) Bi-layer nanostructures of CuPc and Pd for resistance-type and SAW-type hydrogen gas sensors. Sens Actuators B. https://doi.org/10.1016/j.snb.2012.10.056
13. Jakubik W, Powroźnik P, Wrotniak J, Krzywiecki M (2016) Theoretical analysis of acousto-electrical sensitivity in SAW gas sensors with single and bi-layer structures. Sens Actuators B 236:1069–1074
14. Jakubik WP, Urbańczyk MW, Kochowski S, Bodzenta J (2002) Bilayer structure for hydrogen detection in a surface acoustic wave sensor system. Sens Actuators B 82(2–3):265–271
15. Jakubik WP, Urbańczyk MW, Kochowski S, Bodzenta J (2003) Palladium and phthalocyanine bilayer films for hydrogen detection in a surface acoustic wave sensor system. Sens Actuators B 96(1–2):321–328
16. Jakubik WP, Urbańczyk MW, Bodzenta J, Pietrzyk MA (2005) Investigations on the resistance of the metal-free phthalocyanine and palladium bilayer sensor structure influenced by hydrogen. Sens Actuator B 105(2):340–345
17. Jakubik WP, Urbańczyk MW, Bethanabotla V, Cular S (2006) Acoustoelectric effect in Surface Acoustic Wave gas sensors (SAW) with metal-free phthalocyanine and palladium sensing bi-layers. ECS Trans 1(19):1–9
18. Jakubik WP, Urbańczyk MW, Maciak E (2006) Palladium and metal-free phthalocyanine bilayer structures for hydrogen detection in the SAW sensor system based on interaction speed. IEEE Sens J 6(5):1178–1185

19. Jakubik W, Urbańczyk M, Maciak E (2007) Metal-free phthalocyanine and palladium sensor structure with a polyethylene membrane for hydrogen detection in SAW systems. Sens Actuators B 127(1):295–303
20. Jakubik WP, Urbańczyk M, Maciak E, Pustelny T (2008) Surface acoustic wave hydrogen gas sensor based on layered structure of palladium/metal-free phthalocyanine. Bull Pol Acad Sci Tech 56(2):133–138
21. Jakubik W, Urbańczyk M, Maciak E, Krzywiecki M (2011) Bi-layer nano-structures of CuPc and Pd in SAW and resistance hydrogen sensor. Procedia Eng 25:252–255
22. Jakubik WP, Urbańczyk M, Maciak E, Pustelny T (2009) Bilayer structures of NiO_x and Pd in surface acoustic wave and electrical gas sensor systems. Acta Phys Pol A 116(3):315–320
23. Jakubik W, Wrotniak J, Powroźnik P (2018) Theoretical analysis of a surface acoustic wave gas sensor mechanism using electrical conductive bi-layer nanostructures. Sens Actuators B 262:947–952
24. Jakubik W, Wrotniak J, Kaźmierczak-Bałata A, Stolarczyk A, Powroźnik P (2023) Light-activated SAW sensor structures with photoconductive polymer films for DMMP detection. Sens Actuators B 384:133597. https://doi.org/10.1016/j.snb.2023.133597

Index

A

Acoustoelectric (AE), 1–4, 6, 9, 12, 18–21, 26–28, 33, 44, 45, 47, 48, 51, 53–55, 57, 62, 64, 66–68, 75, 77, 79–84, 86–88, 93, 94, 106, 114, 117, 119, 122, 127–129, 131, 132, 134, 136, 137, 140, 141, 152–154, 157, 159–161, 172, 175–177, 181–183

Acoustoelectric parameter of single structure (ξ), 67, 153

Acoustoelectric parameter of the sensor structure under the light activation/ under the light with specific wavelength ($\xi(\lambda)$), 162, 172, 175, 176

Acoustoelectric sensitivity of the single structure/or under the light activation (S_{AE}), 82, 83, 85

Activation/under the light with specific wavelength ($\xi(\lambda)$), 162, 166, 175

Additional external acceleration (a_z), 6

Amplifier phase shift (ϕ_A), 17

Amplitude of oscillator signal, wave amplitude, amplifier symbol (A), 94

Aperture of interdigital transducer, width of the acoustic beam (A), 13

Average flux of wave power per unit of sound beam width (P), 56

Average surface mass density of the structure (ρ_{srf}), 61

B

Boltzmann constant (k_B), 39

Bulk Acoustic Wave (BAW), 2, 24, 25, 54

C

Capacity per unit length of pair of interdigital transducers (C_s), 27

Carbon Nanotubes (CNT), 21, 28, 35

Change in entropy due to gas sorption (ΔS_g), 38

Change in surface electrical conductivity ($\Delta \sigma_s$), 49

Change in the entropy as a result of gas sorption (ΔH_g), 38

Change in the surface density of the sensor element (Δm), 9

Change in the surface mass density of the sensor element ($\Delta \rho$), 9, 47

Change of the combined SAW propagation constant (Δk), 54, 55, 57, 69

Charge Coupled Devices (CCD), 11

Chemical Vapor Deposition (CVD), 11

Circular frequency (pulsation) of undisturbed wave (ω_0), 18

Coefficient related to the relative rate of adsorption and desorption (***B***), 39

Combined constant of SAW propagation after the sensor structure is made (k'), 54, 59, 69

Combined SAW propagation constant without sensor structure (k), 54, 55, 69

Complementary Metal-Oxide-Semiconductor (CMOS), 11

Components of deformation tensor (S_{jz}), 59

Components of the mechanical stress tensor (T_{ij}), 56

Component vector of electric induction (D_j), 56

W. Jakubik, *Application of Surface Acoustic Waves in Gas Sensors*, Springer Series in Surface Sciences 69, https://doi.org/10.1007/978-3-031-89971-3

Concentration of substance in the sensor element (c_{se}), 37, 50
Concentration of vapor or gas in the surrounding atmosphere (c), 37
Constant depending on the parameters of the piezoelectric substrate material (A_p), 58, 63
Constant diffusion of electric charges (D_e), 65, 66, 68, 87
Constant propagation of partial waves generated to the sensor structure (β_j), 63
Coupling parameters between SAW and piezosubstrate (c_j), 58, 60
Coverage width of the electrode system in the planar electrical method (l), 106, 172
Current density in the semiconductor layer (J), 65, 70

D
Debye shielding length (λ_D), 65
Degree of coverage of sensor element (θ), 39, 46
Delay line coverage factor (κ), 62, 99, 100, 119
Density of substrate (ρ_p), 27
Density of surface active centers (absorption states) (N_s), 39, 46, 62
Design parameter determining the electrical conductivity ratio of the bilayers (x), 133, 154
Difference in electric potentials as a result of dipole layer production (, 46
Differential frequency (Δf), 43, 51, 95, 97, 98, 100, 108, 109, 111, 113, 118–120, 122–125, 127, 131, 134, 135, 141, 182
Dimethylomethylophosphor (DMMP), 2, 4, 25, 26, 28, 157, 158, 162–171, 173–177, 182, 183
Dipole moment of adsorbed hydrogen atom (μ_H), 47
Direction of the SAW propagation (z), 18
Distance from the surface of the SAW propagation (y), 7
Dynamic viscosity of the viscoelastic structure (η), 62

E
Effective electrical permeability of the piezoelectric substrate at a constant stress field T (ε_p^T), 65
Elastic modules for acoustically thick structures (M_j), 63
Elastic modulus of the sensor element (C_{11} C_{44}), 47
Electrical conductivity of sensor structures obtained from planar measurements (σ_{rs}), 86
Electrical conductivity of the sensor layer (σ), 3
Electrical impedance (Z_0), 65, 72
Electrical permeability of the substrate (ε_p), 66
Electrical surface conductivity of the polymer film calculated from the sample resistance ($R_s(light)$) after the light activation ($\sigma_{s(light)}$), 172
Electrical surface conductivity of the sensor layer (σ_s), 67, 93, 102, 119, 132, 154, 159
Electric field strength (E), 47, 70
Electric potential (φ), 7, 14, 56, 64–66, 159, 175
Electromechanical coupling coefficient (K^2), 12, 13, 19, 27, 54, 66, 67, 72, 88, 114
External flow (Q_z), 6
External mechanical stress (T_{mz}), 6
External pressure (p_z), 6, 72
External temperature (T_z), 6

F
Field of displacement of substrate particles around equilibrium positions ($\boldsymbol{u}$), 18
Flexural Plate Wave (FPW), 22, 25
Frequency of oscillation on the free surface of the substrate (without cover) (f_0), 16

G
Gas concentration in the measuring chamber at steady state (c_0), 110
Gas constant (R), 38
Gas flow through the measuring chamber (Q), 100, 110, 163

H

High Frequency (HF), 13, 27, 28, 113, 165, 167
Human Immunoglobulin type G (HIgG), 22
Hydrogen diffusion ratio to palladium volume (D_0), 45

I

Imaginary part of sensor element material shearing module (G''), 62
Imaginary part of the volumetric module of the sensor layer (K''), 62
Initial surface electrical conductivity of the structure (σ_{s0}), 49
Initial surface mass density of the sensor structure (ρ_0), 47
Intensity of light illumination (I), 157, 159, 162, 164, 169, 176, 178
Interdigital Transducer (IDT), 2, 3, 5, 13–16, 19, 20, 22, 23, 25, 95, 98, 100, 183
Inverse Mode Configuration (IMC), 114, 134

K

Kind of Photoconductive Polymer (SilPEG1.4), 169

L

Lame constant of the isotropic sensor layer (λ'), 61
Lame constant or isotropic sensor layer (μ'), 61, 119
Length of sensor structure (L_s), 99
Length of the delay line (L), 62, 98
Light-Activated Sensor Structures (LASS), 157–159, 161, 164, 172, 177
Light Activation (LA), 2, 4, 26, 28, 157–159, 161, 164, 166, 167, 169, 172, 174–178, 182
Light Emitting Diode (LED), 4, 133, 157, 164–176
Limit Of Detection (LOD), 151, 170, 171
Lithium Niobate ($LiNbO_3$), 3, 11, 13, 14, 20, 47, 58, 61, 63, 66, 67, 70, 81, 82, 94, 111–114, 117–119, 121, 124, 125, 129, 130, 132

M

Mass of a gas molecule (m), 39
Mass sensitivity in the oscillator system (S_m^f), 9
Mass sensitivity (S_m), 1, 2, 9, 19, 22–24, 28, 62, 80, 159
Measuring chamber filling time constant (τ_K), 110
Metalophthalocyanines (MPc), 40, 42, 43, 46, 64, 67, 117, 137, 160, 182
Metal-Oxide-Semiconductor (MOS), 44, 46, 117, 137
Micro-Electromechanical Systems (MEMS), 11
Microwave Integrated Circuits (MIC), 11
Mobility of electric charge carriers (μ), 65
Multi-Walled Carbon Nanotubes (MWCNT), 35

N

Normalized electrical conductivity of the second layer (ξ_2), 79, 83–86, 88
Normalized electrical surface impedance of the layer (z_E'), 65
Normal Mode Configuration (NMC), 101–103, 111, 121, 127, 140, 141
Number of electrode pairs of interdigital transducer (N), 13, 106, 172
Number of SAW wavelengths between transceiver and receiver (n_0), 16, 98
Number of SAW wavelengths located along the length of the structure (n), 99

O

Owlstone Vapor Generator (OVG), 162, 163, 167

P

Palladium (Pd), 3, 6, 20, 25, 28, 35, 40, 43–48, 50, 63, 86, 106, 112, 114, 117–126, 128, 129, 131–133, 135, 137, 142, 144, 148–150, 153–155, 160, 181, 183
Parameter defining the boundaries between acoustically thin and thick structures (R^*), 58, 63
Parameters of the Sensor Element (PSE), 9–11, 18, 19, 34–36, 39, 40, 52, 53, 94
Particle pressure of gas particles in the surrounding atmosphere (p'), 39
Partition coefficient (K), 37

Phase speed of SAW propagation (v), 9, 34, 66
Photoconductive properties (PC), 157, 174, 177
Physical Vapor Deposition (PVD), 11
Planck constant (h'), 39
Plasma Polymerized Allylamine (PPAa), 22

Q
Quartz Crystal Microbalance (QCM), 13, 25, 26, 54

R
Real part of sensor element material shearing module (G'), 63
Real part of the sensor layer volumetric module (K'), 62
Regioregular Poly-3-Hexylothiophene (rrP3HT), 2, 4, 28, 157, 164–170, 172–177, 183
Relative Humidity (RH), 41, 113, 135, 136, 139, 140, 144, 145, 162, 170, 171, 182
Relative Sensitivity Increase Parameter (RSI), 165, 167, 168
Residence time (gas molecules remaining on the surface of the structure) (τ_r), 38

S
Sample electrical resistance determined from measurements (R_{sam}), 106, 133
Sample electrical resistance under light activation ($R_s(light)$), 172, 174, 176
Sample resistance after deposition (R_{s0}), 172, 175, 176
SAW damping change ($\Delta\alpha$), 160
SAW damping on free substrate or angle of light illumination (α), 53
SAW damping on substrate with sensor structure (α'), 53
SAW speed after the sensor element manufacturing (v'_0), 53
SAW speed change (Δv), 47, 55, 75, 79
SAW speed on the free surface of the piezoelectric crystal (v_0), 47, 53, 55, 59, 71
SAW wave number (k_0), 7, 18, 54, 55, 67, 69–71, 73, 74, 160
Semiconductor electrical permeability (ε), 87
Sensitivity on the number of photons ($S_{\Phi N}$), 169
Shear-Horizontal Acoustic Plate Mode (SH-APM), 17
Signal delay time on free surface of substrate (τ_0), 16
Single-Phase Unidirectional Transducer (SPUDT), 14, 28
Single transducer phase shift (ϕ_{tr}), 16, 17, 98
Single-Walled Carbon Nanotube (SWCNT), 35
Spatial density of electric charges (ρ_e), 65, 69
Standardized electrical conductivity of the first layer in bilayer structure (ξ_1), 75, 79, 82, 88
Strain modulus taking into account the relative stress-to-strain ratio ($E^{(j)}$), 59, 60, 62
Strength of the external electric field (E_z), 6
Structure Work Point (SWP), 159–161, 172, 175, 177
Substrate thickness (d), 27
Surface Acoustic Waves (SAW), 1–20, 23–25, 27, 28, 33–37, 42–48, 50, 51, 53–58, 60, 62–77, 79–81, 86–88, 93–104, 111, 113, 114, 117–125, 129, 132–135, 138, 141, 147, 152, 154, 155, 157–162, 164, 165, 167, 170–172, 174–178, 181–184
Surface density of electric charges (ρ_s), 62
Surface mechanical impedance of the sensor structure (Z_{mj}), 59, 60
Surface Skimming Bulk Wave (SSBW), 22
Surface Transverse Wave (STW), 17, 21–25, 27, 54

T
Temperature of action of the sensor structure (T), 38
Thickness of the first film in bilayer structure ($h1$), 68, 76–79, 88
Thickness of the sensor layer (h), 7, 20
Time (t), 1, 3, 12, 15, 17–22, 26, 28, 36–39, 42, 43, 45, 50, 54, 56, 58, 75–77, 83, 87, 97, 109–111, 114, 121, 123–125, 127, 128, 133, 137, 139, 141, 142, 148, 154, 158, 162, 165–167, 170, 171, 176, 178, 181–183
Time delay of acoustic line (τ), 15
Total electric potential (φ_c), 70

Total wave energy density per unit area (u_C), 72
Tungsten Oxide (WO_3), 3, 6, 21, 42, 49, 113, 117, 122, 124, 137, 138, 142–147, 149–155, 160, 181–183

U
Ultraviolet radiation (UV), 175

V
Vacuum electric permeability (ε_0), 46
Velocity components of points on the surface of the substrate (v_j), 56, 59

W
Wave circular frequency (ω), 18
Wave frequency, line frequency with sensor structure (f), 14
Wavelength of the SAW or light illumination (λ), 7
Width of the electrode and gap of the interdigital transducer (p), 13, 106

MIX
Papier aus verantwortungsvollen Quellen
Paper from responsible sources
FSC® C105338

If you have any concerns about our products,
you can contact us on
ProductSafety@springernature.com

In case Publisher is established outside the EU,
the EU authorized representative is:
Springer Nature Customer Service Center GmbH
Europaplatz 3, 69115 Heidelberg, Germany

Printed by Libri Plureos GmbH
in Hamburg, Germany